Artificial Intelligence, Cybersecurity and Cyber Defense

Cybersecurity Set

coordinated by
Daniel Ventre

Artificial Intelligence, Cybersecurity and Cyber Defense

Daniel Ventre

iSTE

WILEY

First published 2020 in Great Britain and the United States by ISTE Ltd and John Wiley & Sons, Inc.

ISTE Ltd
27-37 St George's Road
London SW19 4EU
UK

www.iste.co.uk

John Wiley & Sons, Inc.
111 River Street
Hoboken, NJ 07030
USA

www.wiley.com

Library of Congress Control Number: 2020940262

British Library Cataloguing-in-Publication Data
A CIP record for this book is available from the British Library
ISBN 978-1-78630-467-4

Contents

Introduction

Cyberspace, from the laying of its first building blocks in the 1950s (the first computers, the first software), to what it has become today – a vast, extremely dense network made up of billions of computers, electronic chips, data flows, with billions of users increasingly dependent on this technological environment – has continued to grow and expand, in a movement of expansion that nothing seems to be able to stop. It has transformed the world, to the point where it has reached the status of a new dimension, alongside the land, the sea, the air and space. Although there was no mention of "cyberspace" in the mid-20th Century, the foundations had already been laid.

Since then, this expansion has been motivated as much by scientific as by economic and industrial motives, as well as by political issues.

The recent technological markers of this development are, for example, "Big Data", "cloud computing", "4G", "5G", the "Internet of Things", "IPv6" and "social media". At the political and societal level, the manifestations of this expansion can be seen by access to the Internet, which affects an ever-growing population, by increasingly "digital" societies (e-administrations, social media, etc.) and by economies that are increasingly dependent on networks, the Internet, computers and communication tools. Some see the effects of this expansion of cyberspace as the democratization of societies, greater freedom of expression and a more dynamic and efficient economy, while others deplore the restriction of freedoms and the strengthening of state control and surveillance capacities, and therefore see this as a cause for the destabilization of societies.

In this landscape, the concept of artificial intelligence (AI) (re)asserted itself around 2015. AI is an element of cyberspace, one of the driving forces behind this broad movement of continuous expansion. But even before thinking about its transformative capabilities – will AI have such a profound impact on cyberspace that it will change its nature? – it is worth observing how AI fits into cyberspace, alongside all the pre-existing bricks and mortar. Before we think of "transformation" or even "revolution", we must think of "contribution" and "adaptation".

AI is "cyber" in its own right, because it exists only through software, programming languages, computers, microchips and the data it feeds on and produces. AI is a scientific discipline and a set of techniques that originated at the same time as computers, at the end of World War II, in the United States, before spreading to the rest of the world. The histories of AI, computers, computing and cyberspace will continue to be closely linked.

Paradoxically, however, today's treatment of AI gives the impression that there are two distinct concepts. Articles, books, colloquia, organizations, doctrines and defense strategies, to name but a few, seem to separate AI and "cyber", with one exception: when AI is understood as a set of new techniques in the field of cybersecurity. In our view, however, AI should be entirely included in cyberspace, of which it is only one of the bricks. It should therefore be seen as a component that will be part of a pre-existing whole, on which it will have an impact, but whose nature and possibilities will also be constrained by this pre-existing environment:

> "No doubt you have been deluged with a flood of reports about AI/ES[1]. Almost simultaneously, you have read that it will revolutionize the way you do business [...] If you are like most members of senior management whom we have surveyed, you are intrigued by the possibilities of this powerful new tool but are too busy, intimidated, or simply unable to discern the truth. But you should, because the bottom line implications are staggering." [GIL 85]

1 AI/ES: Artificial Intelligence/Expert Systems.

It was in these terms that Temple University invited managers from the business and administrative sectors to attend one of its conferences dedicated to AI in 1985. Today, AI still raises many questions, the same curiosity, and its transformative power is still highlighted, with the difference, however, that it is no longer a question of limiting the revolution to a few areas. It will affect all areas of human activity.

AI is already present, everywhere around us and in our daily lives, without us being aware of it most of the time. Modern vehicles that are equipped with electronics and computer technology are equipped with AI, where the software analyzes data from thousands of sensors in real time and performs calculations to regulate fuel supply and stabilize the vehicle, as well as many other functions. In autonomous vehicles, information systems take control: AI analyzes the data and "decides" the actions to be taken to drive the vehicle. Mobile phones, virtual assistants, social networks, anti-spam filters, video games, search engines, medical diagnostic support systems and simulation platforms are some of the applications that have benefited from AI, though it is impossible to draw up an exhaustive list. AI applications multiplied, in particular, during the 1980s, in the form of expert systems, of which industry, as well as defense forces, were major consumers.

The current craze for AI is part of a series of repeated cycles, so there is no guarantee that it will last. However, for the time being, States have been won over, seeing it as a new means of achieving their ambitions: economic and industrial renewal, a new society, governance, AI as an instrument of power and might, and so on.

In this book, we will try to understand how AI has gradually been integrated into defense policies, strategies and doctrines, and, more specifically, into cybersecurity and cyber defense. The integration of AI into the fields of security and defense is the result of a long process which has its roots in the 1950s. Current debates on the use of AI in military affairs necessarily refer to debates on RMA (Revolution in Military Affairs), on changes in the institution of the military, its organization, doctrines and challenges, on changes in power relations, on weapons (arms race, militarization, control), on the law of armed conflict and on the form and nature of conflicts.

This book is structured with chapters dealing with the history of AI (history of research, with which armies are closely associated, in some states), definitions of concepts (in what terms is AI mentioned? How does the military define AI and integrate it into its discourse?), policies (national cybersecurity strategies) and AI in military affairs (military strategies for AI, concepts, doctrines, cybersecurity and cyber defense, the different players). This book is a synthesis of current debates on the role of AI in defense affairs. Since AI is "cyber" by nature, this book explores one dimension of the militarization of cyberspace. Is this renewed interest in AI the starting point for a reconfiguration of cyberspace, and for a new way of thinking about cybersecurity and cyber defense issues, cyber conflict and its multiple forms and manifestations? Are we moving towards a reconsideration of convictions hitherto held about the specific characteristics of cyberspace, such as its fluidity, the impossibility of establishing borders and the difficulty of marking or defending sovereignty, the impossibility of allocating actions, and the difficulty of controlling information?

In this book, the reader will find numerous bibliographical references, (sometimes long) quotations and several summary tables, as we wanted to create a useful working tool for further developments.

	Cyberspace		
L3	No AI	AI	Psycho-cognitive. Essential dimension for AI, because the main challenge of AI is to reproduce and imitate cognitive capacities. Creation of fake images, fake videos, fake Internet users, manipulation of information, perceptions, etc.
L2	No AI	AI	Software, applications. Examples: AI-based applications (natural language, image processing, etc.), AI-specific programming languages (LISP, etc.).
L1	No AI	AI	Hardware. Examples: specific chips for AI applications, robots embedded with AI, etc.

Table I.1. *The three strata of cyberspace (L = layer) and the place of AI in each of them*

Armies and armed conflict have integrated computers, computing, networks and the "cyber" dimension. But this kind of computer science is different from what AI suggests or what it already shows. The computer, as Ted Nelson wrote in 1978, is "a box that follows a plan"[2] (meaning that, in

2 [NEL 78]

this concept, the computer executes code, line by line, as programmed, and cannot go beyond it). AI promises precisely to break with this logic, it promises surprises and new worlds. Is the prospect of machines that can make their own decisions, formulate their own plans and "think" something that army commanders can dream of at the beginning of this 21st Century?

1

On the Origins of Artificial Intelligence

1.1. The birth of artificial intelligence (AI)

1.1.1. *The 1950s–1970s in the United States*

Alan Turing's article, published in 1950 [TUR 50], which is one of the founding works in the field of AI, begins with these words: "I propose to consider the question, 'Can machines think?'"

In 1955, "Logic Theorist", considered to be the first AI program, was developed. This work was the result of cooperation between three researchers: a computer scientist (John Shaw) and two researchers from the humanities and social sciences (Herbert Simon and Allen Newell) [SIM 76]. The application was programmed using IPL language [STE 63], created within the RAND and the Carnegie Institute of Technology[1] (a project that received funding from the US Air Force). Here we have the essential elements of AI research: a multidisciplinary approach, bringing together humanities and technology, a university investment and the presence of the military. It is important to note that although the program is described today as the first AI code, these three researchers never use the expression "artificial intelligence" or present their software as falling into this category. The expression "artificial intelligence" appeared in 1956, during a series of seminars organized at Dartmouth College by John McCarthy (Dartmouth College), Claude Shannon (Bell Telephone Laboratories), Marvin Minsky (Harvard University) and Nathaniel Rochester

1 Today, Carnegie Mellon University. In 1956 the Carnegie Institute of Technology hosted its first IBM computer. However, it was not until 1986, 30 years later, that the university established the School of Computer Science.

(IBM Corporation). The aim of this scientific event was to bring together a dozen or so researchers with the ambition of giving machines the ability to perform intelligent tasks and to program them to imitate human thought.

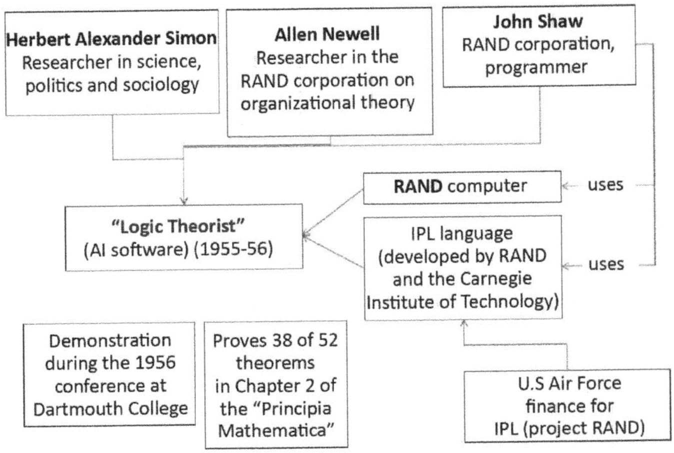

Figure 1.1. *The first artificial intelligence computer program "Logic Theorist", its designers, its results*

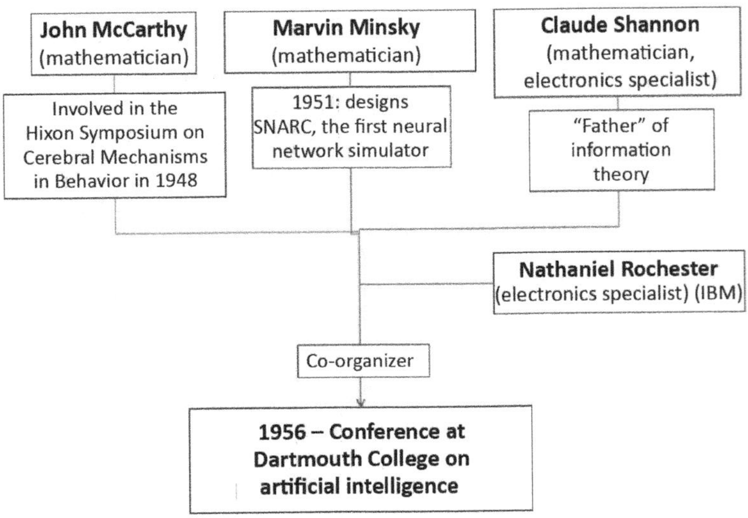

Figure 1.2. *The organizers of a "conference" (two-month program) at Dartmouth College on artificial intelligence in 1956*

While the 1956 conference was a key moment in AI history, it was itself the result of earlier reflections by key players. McCarthy had attended the 1948 *Symposium on Cerebral Mechanisms in Behavior*, attended by Claude Shannon, Alan Turing and Karl Lashley, among others. This multidisciplinary symposium (mathematicians, psychologists, etc.) introduced discussions on the comparison between the brain and the computer. The introduction of the term "artificial intelligence" in 1956 was therefore the result of reflections that had matured over several years.

The text of the proposal for the "conference" of 1956[2], dated August 31, 1955, submitted for financial support from the Rockefeller Foundation for organizing the event, defines the content of the project and the very concept of artificial intelligence:

> "The study is to proceed on the basis of the conjecture that every aspect of learning or any other feature of intelligence can in principle be so precisely described that a machine can be made to simulate it. An attempt will be made to find how to make machines use language, form abstractions and concepts, solve kinds of problems now reserved for humans, and improve themselves."

The project was more successful than expected because 10 people did not participate, but 43 (not including the four organizers)[3], including Herbert Simon and John Nash. This audience was composed almost entirely of North Americans (United States, Canada), and two British people. In any case, it was entirely Anglophone.

In an article titled "Steps toward artificial intelligence" [MIN 61], Marvin Minsky described, in 1961, these early days of AI research and its main objectives:

> "Our visitor[4] might remain puzzled if he set out to find, and judge these monsters for himself. For he would find only a few machines (mostly 'general-purpose' computers, programmed

2 Full document available at: http://www-formal.stanford.edu/jmc/history/dartmouth /dartmouth.html.

3 List published at http://www-formal.stanford.edu/jmc/history/dartmouth/dartmouth.html.

4 Marvin Minsky imagines an alien visitor arriving on Earth for the first time and discovering our computers.

> for the moment to behave according to certain specifications) doing things that might claim any real intellectual status. Some would be proving mathematical theorems of rather undistinguished character. A few machines might be playing certain games, occasionally defeating their designers. Some might be distinguishing between hand-printed letters. Is this enough to justify so much interest, let alone deep concern? I believe that it is; that we are on the threshold of an era that will be strongly influenced, and quite possibly dominated, by intelligent problem-solving machines. But our purpose is not to guess about what the future may bring; it is only to try to describe and explain what seem now to be our first steps toward the construction of 'artificial intelligence.'"

AI research is structured around new laboratories created in major universities. Stanford University created its AI laboratory in 1963. At MIT, AI was handled within the MAC project (Project on Mathematics and Computation), also created in 1963 with significant funding from ARPA.

From the very first years of its existence, the Stanford AI lab has had a defense perspective in its research. The ARPA, an agency of the US Department of Defense (DoD), subsidized the work through numerous programs. The research topics were therefore influenced by military needs, as in the case of Monte D. Callero's thesis on "An adaptive command and control system utilizing heuristic learning processes" (1967), which aimed to develop an automated decision tool for the real-time allocation of defense missiles during armed conflicts. The researcher had to model a missile defense environment and build a decision system to improve its performance based on the experiment [EAR 73]. The influence of the defense agency grew over the years. By June 1973, the AI laboratory had 128 staff, two-thirds of whom were supported by ARPA [EAR 73].

This proximity to the defense department did not, however, condition all its work. In the 1971 semi-annual report [RAP 71] on AI research and applications, Stanford University described its prospects as follows:

> "This field deals with the development of automatic systems, usually including general-purpose digital computers, that are able to carry out tasks normally considered to require human intelligence. Such systems would be capable of sensing the

> physical environment, solving problems, conceiving and executing plans, and improving their behavior with experience. Success in this research will lead to machines that could replace men in a variety of dangerous jobs or hostile environments, and therefore would have wide applicability for Government and industrial use."

Research at MIT in the 1970s, although funded by the military, also remained broad in its scope. Presenting their work to ARPA in 1973, the researchers felt that they had reached a milestone that allowed them to envisage real applications of the theoretical work carried out until then. But these applications cannot be reduced to the field of defense alone:

> "The results of a decade of work on Artificial Intelligence have brought us to the threshold of a new phase of knowledge-based programming – in which we can design computer systems that (1) react reasonably to significantly complicated situations and (2) perhaps more important for the future – interact intelligently with their operators when they encounter limitations, bugs, or insufficient information."[5]

A few leads for new lines of research are then rejected:

> "We believe that AI research can show the way to computer-based information systems far more capable than have ever been available. We plan to attack the area of Information Retrieval, both for traditional data-base and library problems and for more personal management information systems problems. The new Information Systems should help to increase the effectiveness of individuals who are responsible for complicated administrative structures, as well as for complex information problems of technical kinds. In particular, the services will be available and designed to be useable over the ARPANET, and will be designed to interact with the personal

5 MIT, Artificial Intelligence Laboratory, Proposal for research on intelligent automata and micro-automation 1974–1976, Massachusetts Institute of Technology research program supported in part by the Advanced Research Projects Agency of the Department of Defense and monitored by the Office of Naval Research under Contracts N00014-70-A-0362-0003 and N00014-70-A-0362-0005, available at: http://people.csail.mit.edu/bkph/AIM/AIM-299-OCR-OPT.pdf.

systems of other individuals to recognize conflicts, and arrange communication about common concerns."

Along with university teams, the RAND Corporation also played a central role in the emergence of AI in the United States. Willis H. Ware's book, *RAND and the Information Evolution* [WAR 08], is an invaluable resource for understanding the role the organization played in the development of AI in the United States as early as the 1960s. The book covers the period 1946–2008, which is divided into two periods, one from 1946 to 1983, during which research within the agency was organized into departments, before RAND reorganized itself around programmatic actions.

In 1963, Ware recalls, of the 20 contributions that comprise the first book on AI, published by Feigenbaum and Feldman, six were written by RAND researchers. The initial work of Allen Newell and Cliff Shaw, both RAND scholars, in collaboration with Herbert Simon (Carnegie Institute of Technology) laid several foundations for AI research on learning, proof of theories and knowledge representation, among other ideas. Ware also reminds us that AI work does not develop in isolation. Their researchers built on advances in computer science, including intensive uses of new computers such as JOHNNIAC, until the mid-1960s. AI research draws on advances in computer science and many other disciplines.

The history of AI may seem America-centric. But we cannot forget that work in robotics and computer science, in an attempt to understand how the brain works, all converging on AI, mobilized researchers well beyond the North American sphere at the same time.

In the United Kingdom, Grey Walter's research, as early as the 1940s, became part of this international academic movement which was interested in the modeling of brain processes and the definition of intelligence. Grey Walter designed the "turtles" in Bristol in 1947, considered to be the first autonomous electronic robots (Luce Langevin describes the 1950s as a period when an "electronic animal menagerie" was built) [LAN 69]. These approaches were underpinned by the belief in a strong resemblance between the brain and the machine: "Physiologists admit as a working hypothesis that the functioning of the brain is that of a machine" [ASH 51].

Today's international competition between major powers is not unrelated to the history of research and early development from the 1950s to the 1960s. China and Russia, in particular, did not wait until the 2000s or 2010s to invest in the field of AI research. Their present activity in this area is based on a model which, as in the case of the United States, is several decades old.

1.1.2. *AI research in China*

Artificial intelligence research in China began in the late 1960s [XIN 83].

According to Wang Jieshu's analysis [JIE 18], the history of Chinese AI is closely linked to the history of the Soviet Union and the close relationship between the two countries. In the period 1970–1983, the main areas of research covered a broad spectrum of issues, such as:

– machine translation, a field which, in 1982, includes some of the following achievements:

- development of ECTA (English–Chinese Automatic Translation System) software,

- development of the English–Chinese Title Translation System,

- JF-111, A Universal Machine Translation System;

– natural language understanding;

– theorem proving;

– expert systems;

– robotics.

Nothing in this enumeration really distinguishes the orientation of Chinese research from its Western counterparts. Again, the approach here is multidisciplinary (mathematics, computer science, linguistics, medicine, automation, robotics, aeronautics, etc.).

In the 1980s, China already had a large number of publications, achievements, researchers and universities involved in AI research. Jieshu's article only mentions civil applications, nothing is said about the military's position on this research topic and its investment in universities.

On the basis of this article, we identify a set of universities involved (Table 1.1).

Name of University	City (Province)
Zhongshan University	Guangzhou (Guangdong)
Jilin University	Changchun (Jilin)
Zhejiang University	Hangzhou (Zhejiang)
Nanjing Technology College	Nanjing (Jiangsu)
Beijing Aeronautical Engineering Institute	Beijing
Beijing Academy of Traditional Chinese Medicine	Beijing
Institute of Automation, Academia Sinica	Beijing
Institute of Linguistics, Chinese Academy of Social Sciences	Beijing
Institute of System Science, Academia Sinica	Beijing
Mathematics Institute, Academia Sinica	Beijing
Qinghua University	Beijing
Science-Technology Information Institute and Computer Technology Institute, Academia Sinica	Beijing
Shanghai Institute of Computing Technology	Shanghai
Shenyang Institute of Automation, Academia Sinica	Shenyang (Liaoning)
Wuhan University	Wuhan (Hubei)

Table 1.1. *Universities involved in AI between 1970 and 1983 in China (classified by city). Reconstructed from [XIN 83]*

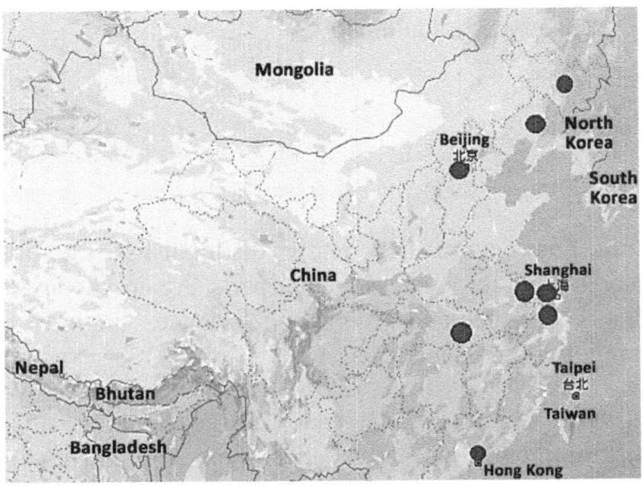

Figure 1.3. *Geographic distribution of Chinese universities investing in AI between 1970 and 1983. Reconstructed from [XIN 83]*

AI research in China thus took shape at the same time as it did in the West and has been structured around universities of excellence. This history serves as a basis for China's current ambitions, which are still expressed in research programs as well as in economic, industrial, societal and political projects.

1.1.3. *AI research in Russia*

The history of AI in Russia follows roughly the same chronology as that of the United States or the West. As early as the early 1960s, Western delegations visiting Russia noticed the presence of research teams on AI themes.

The report by E.A. Feigenbaum, who visited the USSR from June to July 1960 as a member of the American delegation to the *First Congress of the International Federation of Automatic Control* (IFAC) [FEI 60] said:

> "The program consisted of a number of welcoming speeches, and an address by the well-known scientist and Chairman of the USSR National Committee for Automatic Control, V.A. Trapeznikov."
>
> "The Soviet Deputy Premier talked on the problems which automation would bring to 'certain societies' which were not well equipped to handle this kind of technological change – change which would bring unemployment, re-education problems [...]"
>
> "In general, Soviet papers could be characterized as oriented toward theory, while papers of Western delegates mixed theory and application."
>
> "In conjunction with the conference, various research institutes, educational institutions, and plants were officially opened for technical excursions by the delegates [...] By far, the most popular tour was one to the Institute of Automation and Telemechanics in Moscow."

In the Soviet Union, AI was one of the components of cybernetics, in the same way as information theory, computer science or the study of military C2 [LEV 64]. Cybernetics, which appeared in the USSR in 1953, was a new and broad field, organized around various research communities which come together, in particular, at conferences dedicated to cybernetics and in numerous academic publications from the early 1960s. Military cybernetics became a sub-domain of cybernetics.

An article published in the journal *Science* on August 27, 1965 [KOP 65] introduced a new city of science, which had just been built, in Siberia: Akademgorodok, located in the suburbs of Novosibirk (Siberia).

The work of Paul R. Josephson [JOS 97] gives a whole chapter to the history of the birth of AI in the city of Akademgorodok, in the middle of the Soviet period (1960–1970). For it was there that AI in Russia was born. A Russian research community centered on AI was created there in the 1970s, with a university research center, "clubs" ("Akademia" club on Artificial Intelligence)[6] and a Council for AI (the Artificial Intelligence Council)[7], etc.

The city, now considered one of Russia's Silicon Valleys (along with Moscow and St. Petersburg), is said to be home to Russia's "cyberwar soldiers" [CLA 17]. Akademgorodok is home to a technopark, concentrating 24% of the revenue of all Russian technoparks, and 22% of the companies hosted in Russia in technoparks [LOG 16]. The Akademgorodok technocenter is currently reported to host 340 companies, 115 start-ups and nearly 10,000 employees. This ecosystem is complemented by the many university research centers that have made the city famous and unique, due to their high concentration.

In the mid-1970s, the Soviet Union envisaged the use of networked information technology as a tool for controlling, managing and planning for the Soviet economy. The project envisaged at that time was to link major production and political centers using a vast network of computers. Moscow would be the hub, but it would also pass through Leningrad (as St. Petersburg was called at the time) and Kiev. Implementation was to start around 1975 and be fully operational by 1990. Western, American (Control Data Corporation) and British (International Computers, Ltd.) companies were even involved in this project [LIE 73]. The computer and computer science, in the broad sense, was a tool of the Soviet political project, as well as posing a challenge to America which at that point faced difficulties in implementing such networks on a large scale. Soviet technological development was based

6 http://ershov.iis.nsk.su/en/archive/subgroup?nid=763577id_1=763577.

7 http://ershov.iis.nsk.su/en/archive/subgroup?nid=763551id_1=763551.

on a policy of transfer from the United States to the USSR from the end of the 1950s and accelerated from 1972 onwards under the Nixon administration [ROD 81]. The acquisition of foreign technology, especially in the field of information technology, by the USSR, was carried out through legal (sales authorized by the US government) and illegal (black market and copying) channels. AI was part of this Soviet "cybernetics" project. However, a report by the American Department of Defense in 1990 estimated that the Soviet Union had lower capabilities than America, despite research efforts and special attention to AI applications in the civil and military fields [DOD 90]:

> "The Soviet Union lags behind the United States significantly in machine intelligence and robotics. They do have a good theoretical understanding of the area and can show creativity in applying the technology to selected space and military applications. Soviet R&D on artificial intelligence (AI), under the auspices of the Academy of Sciences of the USSR, includes work on machine vision and machine learning. The value of machine intelligence to battlefield operations as well as to the domestic economy has been recognized by the Soviet government."

So, while the Soviet Union does not appear to have been truly competing with US capabilities at the time, it was nonetheless a player that added to the competitive landscape facing the United States. AI and robotics, in their various dimensions (research, development, industrialization), emerged dynamically in several countries and regions of the world: the report cites France, Europe and Japan, among others.

Research in the USSR was not isolated from the rest of the world. The USSR organized international AI conferences: for example, in Leningrad in April 1977[8] and in October 1983[9]. Its research projects and achievements were in fields relatively similar to the rest of the world: applications for automatic translation and understanding natural language (in 1982, the "Etap-1" project was created – Experimental System for Automated Translation from French into Russian)[10].

---

8 http://ershov.iis.nsk.su/en/archive/subgroup?nid=763546id_1=763546.

9 http://ershov.iis.nsk.su/en/archive/subgroup?nid=763550id_1=763550.

10 http://ershov.iis.nsk.su/en/archive/subgroup?nid=763438id_1=763438.

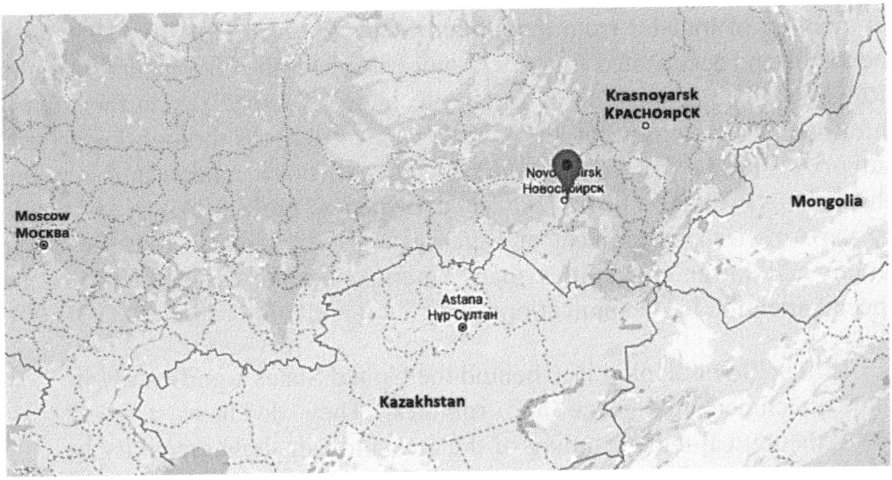

Figure 1.4. *Location of Akademgorodok (district of the city of Novosibirsk)*

1.1.4. *AI research in Japan*

The history of AI research in Japan began in the 1960s at Kyoto University, with a team formed around Professor Toshiyuki Sakai, who worked in three areas: computer vision, speech processing and natural language. In 1970, the team presented the first facial recognition system at the Osaka exhibition. During the decade, several teams took shape, at Kyushu University (around Professor Toshihiko Kurihara, with work on kanji-kana conversion systems), at the University of Osaka (with Professor Kokichi Tanaka on knowledge processing issues), at the University of Tokyo and in corporate laboratories such as NTT. However, the emergence of these various teams did not yet constitute a true Japanese AI research community. This took shape around a community of students at the University of Tokyo, the IAUEO, from which the pioneers of Japanese AI such as Hideyuki Nakashima, Koichi Hori and Hitoshi Matsubara emerged.

The government provided several hundred million dollars in funding for long-term university research programs [DON 77]. In 1971, the government launched the PIPS (Pattern Information Processing System) project, which was funded at $100 million over eight years. The computational requirements necessary to achieve the objectives of PIPS necessitated the development of new electronic chips. This phase was financed by the Japanese government as early as 1973, through a new project.

Other substantial funding was mobilized to support research in image processing, speech and natural language processing.

AI seemed to really take off in the mid-1980s [MIZ 04, NIS 12][11], after Japan launched its fifth-generation computer program, in 1982. The Institute for New Generation Computer Technology (ICOT) was the R&D arm of the national fifth-generation computer science program. Among others, ICOT produced the programming language KLIC and the legal reasoning system HELIC-II. In 1983, the ICOT had less than 50 researchers and the research themes, focused around the central project of designing the world's largest computer by 1990, directly concerned AI (e.g. automatic translation systems, automatic response systems, understanding speech, understanding images, problem-solving systems and logic programming machines, etc.).

In 1986, the Japanese Society for Artificial Intelligence (JSAI) was founded. In 1990, the association created the Pacific Rim International Conference on Artificial Intelligence (PRICAI), which aims to structure AI research in this part of the world, thus complementing or counterbalancing the Western initiatives of the IJCAI and the European Conference on Artificial Intelligence.

Alongside university research, industrial R&D activities have made a major contribution to the development of Japanese artificial intelligence. Major industrial groups set up teams dedicated to AI (NTT, Hitachi, Fujitsu, Sony, Honda, etc., are big names in the industry that invested in this field). Some of their developments received a lot of media coverage, such as AIBO (Sony), ASIMO (Honda), TAKMI – Analysis and Knowledge Mining (IBM Research Tokyo), facial recognition tools and oral translation applications for mobiles (NEC) and the humanoid robot HRP-4C (capable of singing and dancing).

Since the 1980s, the Japanese government has maintained its investment in AI, but international competition is now raging and several major powers are outperforming Japan in terms of numbers of scientific publications: while China published 41,000 articles in the field between 2011 and 2015, the United States published 25,500 and Japan 11,700. This phenomenon has

---

11 This is Toyoaki Nishida's interpretation of AI history. Other analyses consider that the organization of the *International Joint Conference on Artificial Intelligence* (IJCAI) in Tokyo in 1979 was the first milestone in this history (R. Mizoguchi). See [MIZ 04] and [NIS 12].

been repeated in the industrial field: the United States is said to have a thousand companies in the field, while Japan has 200–300[12].

Although Japan has been and remains one of the leaders in robotics and AI applied to this field, it is not only because it shows qualities as an integrator of the multiple aspects required (electronics, electricity, computing, automation, etc.), but also because it is a leader in many of these fields; Japan has particularly been a leader in electronics. Robotics in the 1980s required more skills in electronics, microelectronics and mechanics than in computer science.

1.1.5. *AI research in France*

In the 1950s and 1960s, interest in intelligent machines spread to many countries around the world. France was one of the players in this internationalization of research. For example, we can cite the following:

– Pierre de Latil's reflections on artificial thinking [DEL 53];

– the work of Albert Ducrocq (the son of a soldier, he studied political science and electronics and was later a journalist and essayist, and qualified as a cybernetician) who invented the "electronic fox", an autonomous robot on wheels, and who inspired the achievements in the 1970s of Bruno Lussato, inventor of zebulons, autonomous computerized handling robots. Albert Ducrocq published several works dealing with robots, weapons and AI such as *Les armes de demain* (the weapons of tomorrow) (Berger-Levrault, 1949), *Appareils et cerveaux électroniques* (electronic devices and brains) (Hachette, 1952), *L'ère des robots* (the age of robots) (Julliard, 1953) and *Découverte de la cybernétique* (discovery of cybernetics) (Julliard, 1955). He published many other works before his death in 2001;

– writings on thinking machines by Paul Chauchard [CHA 50], Paul Cossa [COS 54], Louis Couffignal [COU 52], Dominique Dubarle [DUB 48], or on the robot, with the writings of Albert Béguin [BÉG 50].

Questions are tackled from a variety of viewpoints: mathematicians, cyberneticians, philosophers, electronics engineers, etc. Are humans machines or robots? Can the brain be reproduced in a machine? Can the machine think, does it have a soul? What is a machine? Can we reduce the mechanism of

12 List of AI companies in Japan available at: https://www.data-artist.com/en/contents/ai-company-list.html.

thought or the functioning of the brain to algorithms? Are the brain and the body simple mechanics?

Louis Couffignal [COU 52] defined the machine as “an entire set of inanimate, or even, exceptionally, animate beings capable of replacing man in the execution of a set of operations proposed by man”.

He listed the categories of machines: machines that can add and write, machines that can read and choose, calculating machines and thinking machines.

In 1984, Jean-Pierre Laurent [LAU 85], professor at the University of Chambéry, drew up a series of observations on the state of French AI research: identifying research potential was a difficult exercise because of the imprecise perimeter of AI, a research discipline in its own right for some, adding techniques with various applications to others; on the other hand, AI research was dispersed across France – the teams were small and spread over the whole country; AI suffered from a rather negative image in the world of academic research and in political decision-making circles, due to the overly empirical nature of the discipline, which was moving away from purely theoretical research, and the lack of convincing and spectacular results, capable of overcoming the failures of the 1960s. AI was excessively associated with “gadget” applications, but at the beginning of the 1980s, the situation seemed to be changing, and developments in the industry using expert systems seemed likely to give AI the serious image it lacked. Nevertheless, the industry still had only an imprecise vision of what AI could bring: “Some industrial organizations that are not really familiar with AI and its possibilities are inclined to believe in miracles and ask AI to solve everything (how, for example, can an expert system be built if there are no human experts?).” Researchers investing in the discipline were still poorly supported and had very limited means in terms of human resources, equipment and funding. This situation had an impact on university teaching, which attached only minor importance to AI training. The situation of academic research therefore depended on a wide range of variables: public interest, the attention of political decision-makers, industrial investment and the ability to unite isolated researchers and disciplines.

It should also be noted, when reading this study by Jean-Pierre Laurent, that the drivers of research funding depended on exogenous variables. Although the State, via its research organizations, such as the CNRS, had been funding some AI projects since 1979, the publication of the Fifth Generation MITI Project (Japan) in 1981 seemed to be a new decisive moment in political action.

1.2. Characteristics of AI research

From the outset, AI has been formed with a multidisciplinary approach, located at the crossroads of a body of research involving multiple scientific disciplines. Philosophy occupies an important place in this interdisciplinarity (H.B. Dreyfus [DRE 72] establishes links between AI and the thinking of Plato and Hobbes), alongside mathematics, computer science and cognitive sciences in particular.

> "The field draws upon theoretical constructs from a wide variety of disciplines, including mathematics, psychology, linguistics, neurophysiology, computer science, and electronic engineering." [LAW 84]

In his March 2018 report, French MP Cédric Villani recalled this interdisciplinary approach:

> "The field is so broad that it cannot be limited to a specific area of research […] AI is at the crossroad of multiple fields: computer science, mathematics (logics, optimization, analysis, probabilities, linear algebra, etc.), cognitive science, etc." [VIL 18]

Spain, in its official National Strategy Document for AI (2019), does the same: "Due to the growing complexity of its contributions, it is increasingly interdisciplinary, with synergies with biology, philosophy, the world of law, psychology, sociology and economics."[13]

AI uses a wide range of techniques or technologies:

> "each algorithm in AI is supported by a mix of techniques: semantic analysis, symbolic computing, machine learning, exploratory analysis, deep learning and neural networks […]" [VIL 18]

---

13 Spanish RDI Strategy in Artificial Intelligence, Ministry of Science, Innovation and Universities, 2019, available at: http://www.ciencia.gob.es/stfls/MICINN/Ciencia/Ficheros/Estrategia_Inteligencia_Artificial_EN.PDF.

> "These technologies include: natural language processing, smart agents, computer vision, machine learning, expert systems, autonomous cars, chatbots and voice recognition."[14]

> "Six Al technologies have significant potential for application to production and logistics processes: expert/knowledge-based systems, natural language, speech recognition, three-dimensional vision, smart robotics and neural networks. The most significant of these technologies thus far is expert systems." [MEL 89]

> "Technologies and research areas generally considered to be sub-domains of AI: • Automated Planning and Scheduling • Computer Vision • Decision Support, Predictive Analytics, and Analytic Discovery • Distributed Artificial Intelligence/Agent-based Systems • Human Language Technologies • Identity Intelligence • ML • Process Modeling • Robotics/Autonomous Systems." [DNI 19]

Finally, this new science is characterized by the extent of its field of application: in the 1970s, research tried to apply AI to the games of chess and Go, to mathematics (proving theorems), to understanding and translating written and spoken language, to controlling robots and artificial hands, to image analysis and to human–machine interaction (an intelligent dialogue)[15]. In the 1980s, there was talk of applications in:

> "bioengineering, chemistry, computer hardware, computer software, education, engineering, general purpose systems and AI utilities, law, manufacturing and industry, mathematics, medicine, the military, and resource exploration." [LAW 84]

14 Mauritius Artificial Intelligence Strategy, report, A report by the working group on artificial intelligence, November 2018, available at: http://mtci.govmu.org/English/Documents/2018/Launching%20Digital%20Transformation%20Strategy%20191218/Mauritius%20AI%20Strategy%20(7).pdf.

15 *Cryptolog* Magazine, NSA, July 1975, available at: https://www.nsa.gov/Portals/70/documents/news-features/declassified-documents/cryptologs/cryptolog_11.pdf.

Three decades later, the Villani report [VIL 18] evokes an equally broad field of applications: image and video recognition, translation, content recommendation, etc. The Villani report [VIL 18] is a report on the use of the Internet for recognizing images and videos.

The application spectrum seems infinite; the only limits are those of the imagination. Some applications that seemed quite innovative in 2020 are, already in an experimental phase or were planned in the early days of AI:

– AI in cars [TSU 79][16];

– image analysis, detection of objects in images [NAG 79];

– conflict simulators (e.g. CONSIM, *Conflict Simulator*, in 1971 (*Proceeding ACM '71* [CLE 71][17]).

Globally, research and development in AI have the common aim of understanding "how human cognition works by creating cognitive processes that emulate those of human beings" [VIL 18].

16 "This paper describes an automobile with artificial intelligence, which consists of a road pattern recognition unit and a problem solving unit. The vehicle is completely autonomous and can be driven without a human driver. The road pattern recognition unit involving a pair of TV cameras and a processing unit identifies obstacles in front of the vehicle and outputs data regarding the locations of the obstacles. The problem solving unit is a microcomputer system and determines control optimal to the environment around the vehicle based on the data. The algorithm employed in it is a table-look-up method, in which the location of the optimal control is addressed in the table by key words generated from the data. The table was heuristically made by means of digital simulation. The vehicle was successfully driven under various road environments at the speed below 30 Km/h."

17 "The development and use of CONSIM (Conflict Simulator), a computer program designed to heuristically simulate decision making, is described. A typical example is used to evaluate and analyze the methodology of CONSIM. The simulation model is designed for two political opponents, each possessing any finite number of alternatives. The model is constructed utilizing techniques frequently employed in game theory. Probabilities are assigned, using a Bayesian Approach, to sets of alternatives available to the United States and China. Incorporated into the model is the capability of varying probabilities as the Vietnam Conflict evolves and the re-evaluation of risks, costs, and benefits occurs. CONSIM is easy to use and applicable wherever probabilities may be assigned to each alternative in a mutually exclusive and exhaustive set of alternatives in a dynamic situation."

Definition of AI objectives/finalities	Source	Year
"The goal of AI research is to construct computer programs which exhibit behavior that we call 'intelligent behavior' when we observe it in human beings."	[FEI 63]	1963
"The long-term goal of general AI is to create systems that exhibit the flexibility and versatility of human intelligence in a broad range of cognitive domains, including learning, language, perception, reasoning, creativity, and planning."	National Science and Technology Council. Networking and Information Technology Research and Development Subcommittee (NITRD) (United States) [NET 16]	2016

Table 1.2. *The objectives or goals of AI*

Writing a history of AI would require taking into account not just the course of one discipline but the meeting of many. Its birth is a moment of convergence, of encounter. The destiny of AI was and will remain closely linked to this multidisciplinarity which is one of its main characteristics.

1.3. The sequences of AI history

The course of AI history would be punctuated by a succession of advances, phases of mania and periods of setbacks, sometimes called "AI winters". The path of AI is, in any case, not linear. While AI is today a worldwide success (we are in "the third era of artificial intelligence" today [MIA 18]), a new phase of retreat , which could occur rather abruptly cannot be excluded. The success of AI is linked to several conditions. Its destiny does not depend only on the progress of science, but it is linked to economic challenges and investments and to political support as well as to scientific progress.

Dividing the history of AI into several sequences makes it possible to position, within each of them, events, such as the main conflicts that have never ceased to punctuate the history of humanity, political periods and major events, in particular, events in the history of technology.

Period 1956–1970 (Spring 1): during this first period, pioneers of AI were confident in their ability to achieve the goals they had set themselves, assured that the state of science would allow for the rapid realization of AI. These pioneers were optimistic, believing that AI would have achieved its goals within the next 25 years. But difficulties quickly arose and initial ambitions had to be reconsidered. Progress was slower than expected, and many problems were still too difficult to solve for the knowledge of the time. Machine translation systems, for example, did not perform at all well. The Lighthill Report (1972) put an end to this first period for a while. Excessive negative discourse brought about the end of this dynamic period, full of hopes and utopias:

> "ARPA, the Advanced Research Projects Agency of the Defense Department, is in fact the major source of funding for AI research, and is at present under attack from several directions because its projects are not considered to be 'paying off' as rapidly as they should be for the expenditures involved. The more pragmatic of Dreyfus' criticisms, even though they may be dismissed out of hand by AI workers, are echoed by others not so easily brushed aside, for example J. Lighthill (1973) speaking for the Science Research Council in England. If enough critics succeed in discouraging ARPA research, and AI work in general, some of the potential benefits for NSA that might have been around the corner may never materialize."[18]

Period 1970–1982 (Winter 1): in the early 1970s, the so-called "AI winter" began, designating the years during which research had to persevere and obtain significant results while facing a reduction in funding. The efforts of the first period were not in vain, because, in the failures suffered, researchers understood what they had to work on.

In the 1970s, the immediate objective was no longer the creation of a universal, general intelligent machine, but of a specialized intelligence: a particular machine should be able to act intelligently in a limited domain[19]. The almost utopian hopes of AI pioneers, assured of being able to quickly transpose the mysteries of human intelligence into the machine, quickly gave way to more measured debates on the performance of the machine. It was a question of what computers could do or could not do, of their limits [DRE 72].

---

18 *Cryptolog* Journal, NSA, July 1975, available at: https://www.nsa.gov/Portals/70/documents/news-features/declassified-documents/cryptologs/cryptolog_11.pdf.

19 *Cryptolog* Journal, March–April 1986, NSA, available at: https://www.nsa.gov/Portals/70/documents/news-features/declassified-documents/cryptologs/cryptolog_101.pdf.

France appears to have been marked by strong skepticism in the 1970s and 1980s. The State invested little at the time, and the research community was still small:

> "There is a great deal of activity in research in computer vision which, as with speech understanding, is a difficult area, requiring enormous amounts of computer power. Fortunately, the indications are that the power required for many applications will be available and economically feasible in about ten years. Researchers, therefore, are currently attacking parts of the total problem with the presently available resources, leaving the integration of their efforts to the future." [HOL 76]

Period 1982–1987 (Spring 2): Japan's program for a new generation of computers revived the race for AI research, as did Ronald Reagan's so-called "'Star Wars' speech". The great nations embarked on a technological race, so as not to be totally overwhelmed and dominated by Japan. It was in the efforts for AI and the conquest of space that the struggle for power between states on the international scene was manifested. In the 1980s, expert systems proliferated. AI became commercial, and for the military, AI at last offered the prospect of applications to reinforce operational efficiency. Funding was arriving in a massive way.

Period 1987–2010 (Winter 2): the positive phase of the early 1980s was short. The AI market began to show signs of a significant slowdown as early as 1987. The specialized AI hardware industry (LISP machines) collapsed.

Period 2010-now (Spring 3): a new phase of worldwide interest in AI. The period since 2011 onwards can truly be called an "AI Spring"[20]. This period of renewal can be rooted in the conjunction of three variables[21]: Big Data, algorithms and increased computer power. This is a claim that the Finnish authorities repeat:

> "More effective and lower-cost computing capacity, vast increases in the amount of data that can be used by AI, and more advanced AI algorithms have all led to more intensive use

20 National artificial intelligence strategy for Qatar, January 2019, available at: https://qcai.qcri.org/wp-content/uploads/2019/02/National_AI_Strategy_for_Qatar-Blueprint_30Jan2019.pdf.
21 *Idem*.

> of artificial intelligence. In fact, we are experiencing a new spring of artificial intelligence."[22]

These three variables can also be called Big Data, Machine Learning (ML) and Cloud Computing [MIA 17].

Many hopes are once again placed on the progress and contributions of AI, there is massive state and private funding, multi-year research and development programs, and implementation in all sectors of activity where it is possible to do so. Performances well below expectations would undoubtedly be likely to slow down the current global dynamics. This would then result in a drop in funding, a lack of public interest and a decline in the commercial activities of AI products. The general or weak AI that is deployed today carries within it the seeds of its own limitations, and human-like systems, faithfully imitating intellectual capacities, are still out of reach:

> "[…] every aspect of our lives will be changed in some way. Today the progress of these intelligent machines seems limitless. Will they surpass us, or even replace us? If by definition a super-intelligence is capable of surpassing us in all areas of intelligence then our own efforts will become obsolete by then. In some ways these machines have superhuman abilities but in fact there is no machine out there that is as intelligent as a rat."[23]

Further disappointments would lead to a new downturn.

This breakdown of AI history since the 1950s is imperfect, as it reduces all aspects of AI into one. Research cycles are not necessarily the same for industry, military programs, policy or the various application areas. Research has not really gone through a period of hibernation. It has continued to be carried out throughout the world with, of course, support in terms of funding

22 Leading the way into the age of artificial intelligence, Final report of Finland's Artificial Intelligence Programme 2019, Publications of the Ministry of Economic Affairs and Employment, available at: https://julkaisut.valtioneuvosto.fi/bitstream/handle/10024/161688/41_19_Leading%20the%20way%20into%20the%20age%20of%20artificial%20intelligence.pdf?sequence=4.

23 Documentary broadcast on Arte.fr in September 2018, *L'intelligence artificielle va-t-elle nous dépasser?*, directed by Guillain Depardieu and Thibaut Martin. Video available at https://www.arte.tv/fr/videos/069097-000-A/l-intelligence-artificielle-va-t-elle-nous-depasser/.

and resources that vary according to the period and the country. The volume of publications, the efforts of the scientific community to structure itself and organize research around national and international associations and international symposia bear witness to this.

States' efforts in AI R&D are mainly motivated by the desire to build a technological field that provides them with economic power, offering the multiple sectors of activity tools to promote their progress, and in the military field, instruments useful for increasing their capabilities. However, during the decades from 1950 to 2000, there was no open discussion of security applications and issues, even though intelligence agencies were able to use AI to contribute to the accomplishment of their missions.

1.4. The robot and robotics

AI is closely associated with robotics. AI was established as a discipline and a scientific community in the 1950s. Specialist publications, symposiums, associations and the creation of research teams have come to participate in the construction process. AI has integrated the robot into its research questions. But it has not joined forces with robotics, which is a discipline in its own right. Robotics has its own history, its own challenges and its own research questions.

Figure 1.5. *Manchester University Robot*[24]*. For a color version of this figure, see www.iste.co.uk/ventre/artificial.zip*

24 George Arents Collection, The New York Public Library, Manchester University robot, available at: http://digitalcollections.nypl.org/items/510d47da-93ab-a3d9-e040-e00a18064a99. This illustration shows the differential analyzer designed by Vanevar Bush.

In the 18th and 19th Centuries, several achievements made by automatons were presented to the public. These were machines that took on the appearance of human beings and reproduced tasks such as writing or playing an instrument. There was, of course, no autonomous intelligence in these machines. The intelligence was that of their ingenious designers, who created precise mechanics. These machines took their place in the literature and reflected the technologies of the time: not yet called robots, they were formed of electricity, mechanics, steel and sometimes steam energy (see *The Steam Man of the Prairies* by Edward S. Ellis [ELL 68]).

Mechanization would win in the 20th Century; the industry opening the way for large-scale production (e.g. Ford factories in the 1910s). What had until then been automatons, machines and mechanics became "robots". Karel Capek introduced the term in 1921 in a play, designating machines with a human appearance. Isaac Asimov wrote, in 1941, the short story *Liar!* [ASI 41], in which he introduced the "laws" of robotics. In the following decades, science fiction would give a substantial role to the robot and would contribute to its global success. But science fiction also contributed to forging a particular image of the robot: it is often human in appearance, it is, in any case, endowed with life and an intelligence that defies that of human.

During World War II, unmanned aircraft equipped with explosives and thrown at enemy populations and armies were called "robot bombs". After the war, other developments were created with industrial applications. The first programmable arm was designed in 1954 by George Devol and Joe Engleberger, which was integrated into assembly lines at General Motors in 1962. Robots occupied the industrial space, the space of science and were designed to assist humanity in their experiments, and to confront hostile environments. In 1994, the Robotics Institute at Carnegie Mellon University designed the eight-legged robot DANTE II, which explored the crater of the Spurr volcano. Its shape was inspired by the shape of a spider. This achievement combined sensor technology, computer science, electronics and communication instruments.

Robotics, a scientific discipline, came about either in 1956 or 1964:

– 1956, when George Devol and Joe Engelberger met. From their collaboration, the robotics industry, a magazine (*Industrial Robot*) and an international conference (*International Symposium on Industrial Robotics*) were born. Joe Engelberger is sometimes referred to as the "father of robotics";

– 1964, when Bernie Roth, a mechanical engineer, met AI researcher John McCarthy. This was the starting point for a fruitful collaboration, scientific publications, developments and regular scientific meetings that gradually solidified a new research community.

A third option of the date of the creation of this discipline dates back to telerobotics (developed in the 1940s in the United States to manipulate nuclear materials). Goertz's articles in 1952 and 1954 were the starting point for the history of robotics[25].

Whatever the precise date chosen, the birth of robotics is, generally, contemporary to that of AI (1950s).

Robotics, as a discipline, has rubbed elbows with AI since its origins, but it has developed as a discipline in its own right:

> "[...] the scientific task of understanding intelligence is already well under way in artificial intelligence, a subpart of computer science. If an appropriate direct scientific field already exists, why is robotics needed in addition? Why isn't artificial intelligence the supplier of basic scientific capital to robotics, along with other suppliers, such as computer science, mechanical engineering and electrical engineering?
>
> There is an answer and it further illuminates my basic thesis. Artificial intelligence currently shares with computer science a special view – it considers information processing divorced from energetics. This creation of an interior milieu, in which only information processes occur, is a powerful abstraction, one which helped computer science to emerge by permitting it to focus on the essential mechanisms. But the costs of the abstraction show nowhere more clearly than in the unexplored central problem of robotics – controlled perceptually coordinated motion." [NEW 81]

25 These three versions are available in [MAS 12].

The work of the Stanford AI Research Center in the 1970s integrated R&D in robotics:

> "We are pursuing fundamental studies in several areas: problem solving, perception, automated mathematics, and learning. However, we find it productive to choose as a goal the creation of a single integrated system [...] Our system consists of a mobile robot vehicle controlled by radio from a large digital computer. The principal goal is to develop software for the computer that, when used in conjunction with the hardware of the vehicle, will produce a system capable of intelligent behavior [...] Before we changed computers (at the end of 1969), our robot system had achieved a primitive level of capabilities: It could analyze a simple scene in a restricted laboratory environment; plan solutions to certain problems, provided that exactly the correct data were appropriately encoded; and carry out its plans, provided nothing went wrong during execution. Therefore, when we began planning a new software system for controlling the robot from a new computer, we set more difficult short term goals: The system is to be able to operate in a larger environment, consisting of several rooms, corridors, and doorways; its planning ability must be able to select relevant data from a large store of facts about the world and the robot's capabilities; and it must be able to recover gracefully from certain unexpected failures or accumulated errors. We have not yet accomplished these goals." [RAP 71]

Like AI, robotics was built around an American center. However, robotics and robots have met with great success in many other countries, including Japan, a country that is particularly active in research and development in the field: in 1997, the first robot tournament in Japan (RoboCup) took place; in 1999, Sony presented and marketed AIBO, the first pet robot-dog; in 2000, Honda created the humanoid robot ASIMO. Japan has been particularly involved in robotics and has been a world leader in recent decades, ahead of the United States and other industrialized nations. In 2010, Japan accounted for two-thirds of the world's robot production (although not all of this robotics is AI-related). This success is largely due to the fact that Japan has, on its territory, all the technological and industrial skills and building blocks necessary for robotics (without being a leader in all of these fields: mechanics, electronics, computing, sensors, etc.) [NAR 10]. The efforts made in robotization can also be explained by specific societal needs: a lack of manpower and an aging population that no immigration can compensate for. It is therefore necessary to replace humans by machines in production, and to think about solutions to support elderly populations. Robotics is not a priority issue for national defense, especially as, since the end of World War II, the country no longer has the right to have its own army. Investments are therefore oriented in fields other than defense.

Robots are machines (AI is software):

– that are programmed;

– that perform tasks independently or semi-autonomously;

– that interact with their physical environment through sensors.

Some robots are also machines that are controlled by operators (so there is no question in their case of autonomy). The "non-intelligent" robot, as opposed to the robot that is intelligent, is only capable of repetitive tasks: taking and moving an object. For this action, it does not need human intervention, so it is, in a way, "autonomous", but it is difficult to qualify it as "smart". Autonomy is therefore not synonymous with intelligence. However, this robot, which only repeatedly places objects, becomes "smart" when a camera is attached to it, which allows it to distinguish objects, differentiate them and assign a different action according to the type of object. Vision and perception require AI algorithms [OWE 17]. Intelligent robots are robots that have embedded AI programs.

1.5. Example of AI integration: the case of the CIA in the 1980s

Computers entered the agency (the CIA) in the early 1960s, primarily for administrative tasks. Then, the computer emerged as a tool that could be used for the very mission of the agency, i.e. intelligence, and the collection and processing of data[26].

In the 1980s, the computer became an essential part of the agency's activities, and it had to deal with increasing volumes of data that had to be collected, stored, analyzed and distributed within its departments[27]. The

26 https://www.cia.gov/library/readingroom/docs/CIA-RDP90G00993R000100130005-7.pdf.

27 ILO 0751-86, August 29, 1986, declassified, available at: https://www.cia.gov/library/readingroom/docs/CIA-RDP90G00993R000100130005-7.pdf.

agency was confronted with constraints imposed by data, computers and the network: large quantities of data were produced and constant and rapid flows connected machines and individuals. The speed and masses of data were two of the variables put forward by observers of the time. However, the findings did not call the technologies into question, as these seemed to bring only added value to the running of the agency and intelligence officers. The introduction of technology into the agency is portrayed as a dynamic process. In the mid-1980s, AI had not yet been adopted but the agency launched into a process of reflection which mobilized its various services and managers to envisage applications in the intelligence field.

On the other hand, the increase in the speed and density of data production by the agency (in the form of reports, notes and various documents) was not only due to technology. There was a political and governmental will within the agency to produce more knowledge on a wide range of issues every day (800 reports were produced by the CIA in 1983; but half of the staff were focused on the USSR [DDI 84]).

The treatment of AI by the American intelligence community can be analyzed using resources available online. Thus, the CIA offers a set of archival[28] documents on its site. In this corpus of data, we are looking for information that allows us to describe the relationship that the intelligence community has with a science and a set of innovative technologies, the way in which the institution organizes itself to tackle these new objects (creation of groups, designation of leaders, creation of relationships, networks, monitoring, etc.) and the applications envisaged for AI for the specific needs of intelligence, all from a chronological perspective (at what points does the agency take up the subject, does it act, what are the important milestones in this evolution, etc.). This facet of the history of the CIA has long remained unknown and in the shadows:

> "The CIA's role in the application of science and technology to the art of intelligence is far less appreciated […] However, the exploitation of science and technology has been a significant element of the CIA's activities, almost since its creation. In 1962, it resulted in the creation of the Deputy Directorate of

28 Data accessible from: https://www.cia.gov/library/readingroom/. A search on the term "artificial intelligence", using the engine available on the page, directly identifies the corpus useful for our analysis.

> Research, which was succeeded in 1963 by the Deputy Directorate for Science and Technology (renamed the Directorate of Science and Technology in 1965)." [RIC 01]

Moreover, if this story takes into account the CIA's policies of acquisition and ownership of technology, the relationship must also be seen in the opposite direction, that of the influences of intelligence on the academic research community and industry. The CIA cannot be considered as a simple user "client": "It is also responsible for a number of scientific advances" [RIC 01].

1.5.1. *The CIA's instruments and methods for understanding and appropriating AI adapted to its needs*

The observations formulated by the intelligence community, indeed both the defense and the intelligence communities, were strongly marked in the early 1980s by a trend aimed at developing and experimenting with smart computer systems for data processing [ECK 84]. Defense and intelligence officials observed an imbalance between, on the one hand, an ever-growing mass of data and, on the other hand, human resources that are or will inevitably be overwhelmed by the task. The AI perspective emerged as a source of possible solutions to what was a managerial challenge and not yet essentially a security issue. The option of reversing the trend by reducing the volume of data collected, produced or provided to analysts was not considered. The only scenario is that of increasing data volumes, accelerating data flows and overwhelming human capacity with a system of machines that needs to be compensated for:

> "A central objective of Artificial Intelligence is to enable computers to emulate the intellectual functions that humans employ in analyzing and interpreting data. For these reasons the Intelligence Community is interested in assessing the potential of artificial intelligence as a technological key to computer-based smart systems for intelligence applications." [ECK 84]

The particular interest in AI in the early 1980s was part of a special framework, a key moment in the history of the CIA, which partly recovered budgets it had lost over the previous decade. This renewal of means allowed the agency to envisage a new phase of growth, to recruit, to develop its activities, to widen its field of expertise and to increase its productivity and research activities:

> "The last four years have seen remarkable growth in the Intelligence Community's budget [...] it has enabled us to restore many of the capabilities that we lost in 1970s [...] In the last four years, we in the CIA for the first time have developed and implemented a comprehensive research program covering a staggering number of countries and issues [...] we have rebuilt our human intelligence capability [...] On the technical side, investments in new technical systems are beginning to pay off." [DDI 84]

This modernization would increase data collection capacity, which would result in new analytical needs:

> "This means a tremendous increase in the quality and precision of intelligence information we can collect. It also means that the volume of data will substantially increase. This suggests that we must begin to invest in processing systems to match our capability in collecting raw data [...] We have to avoid being overwhelmed by this volume of information from all sources, and that is where developments in computers, software, and other new technologies really must help us." [DDI 84]

Although the obstacle of financial resources seems to have been removed, the implementation of new systems was no less complex. One of the main difficulties lay in defining objectives and the appropriate methods. The low level of competence of the decision-makers (managers within the agency) was also pointed out, with its consequences on the choices made:

> "A serious problem for both American industry and government is the computer illiterate senior manager [...] important decisions on computers and ADP equipment are made by managers who hardly know a mainframe from a Mack truck. What do we do? We turn to the computer specialists in our own organizations who think narrowly and protect their turf and for whom larger scale planning, networking, and experimenting is anathema. And so we sometimes develop inadequate systems that can't talk to each other and meet only today's needs." [DDI 84]

Other particularities must be taken into account, such as the agency's structure and mode of operation, which are reflected in how technological systems are organized and which are sometimes referred to as "problems":

> "The major problems in the use of computers at CIA revolve around compartmentation and security. Unlike organizations of similar size in the private sector, we have to have a system that operates on a need-to-know basis, and that may involve only a handful of people." [DDI 84]

Security imperatives create particular constraints, which the systems must integrate: "We must protect ourselves at the same time against 'hackers' from the outside, and the possibility of 'moles' from the inside" [DDI 84].

The CIA, in the first half of the 1980s, structured its computer system into five independent systems: one system for analysts, one for operations, one for administration, one for personnel security and one for processing and analyzing technical intelligence data. Each system has its own software.

In 1983, the CIA drew up an action plan, the *CIA Strategic Plan* (1983–1993) [FIT 83], which sets out the areas that would specifically be the subject of an intensive development policy: AI is one of these areas, alongside human resources, information handling, space, crisis management, terrorism, secret communications, clandestine information gathering, arms control monitoring and protecting intelligence.

It is in this context of renewal that the agency was interested in AI (at the crossroads of what was then the beginning of the second wave of AI and the fifth generation of computers), in which its managers place significant hopes:

> "We will rely on AI in expert systems applications to enable us to detect indicator anomalies for warning, to synthesize combinations of data for analysis, to scan mail to pick out critical messages, or to pick out gaps in our knowledge. Applications of AI in processing huge quantities of raw data without having to translate raw data into standardized formats, as we now do should help separate the wheat from the chaff, especially in SIGINT and imagery […] More sophisticated simulation and modeling techniques will increase our ability to predict alternative outcomes of future events. AI should help analysts compare dissimilar forms of data – imagery, SIGINT, regular text […] Another application might involve accessing more data on a real time basis, especially in crises." [DDI 84]

There are, however, voices that call for the utmost caution. In the face of the hopes placed in AI's ability to profoundly transform intelligence capabilities, one should be aware of the long road ahead to achieve such achievements:

> "progress in the area of AI, in our view, is likely to be painfully slow. Promises of quick advances with practical applications should be treated with some skepticism." [DDI 84]

The Director of Intelligence (DDI), in a speech to the members of the Security Affairs Support Association, stressed the need for the institution to resist commercial pressures, marketing arguments and promise sellers:

> "I can't tell you how many contractors have tried to sell me software that will enable me flawlessly to predict the next action of the Soviet leadership. Until we can understand more about how the intuitional process works, it seems to me, it will be difficult to write 'expert system software' that can duplicate what analysts do." [DDI 84]

Skepticism is even coupled with a certain irony in the way these new technologies are viewed, at least an awareness of their limitations: "Today's Expert Systems are 'idiot savants', demonstrably useful in strictly circumscribed applications" [ECK 84].

Intelligence expectations of AI in 1983–1984 were quite similar to those of the late 2010s: detection, prediction, raw data processing, image and signal processing, etc.

We can also observe that the AI systems that have been developed since then will not have fulfilled all their promises: expert systems and other modalities for processing multisource data in heterogeneous formats will not have made it possible to predict the end of the USSR, to anticipate and prevent the attacks of September 11, 2001, and many other major actions or events that have since marked the history of both the United States and the world.

The attention paid by both the intelligence community and the military to scientific advances in the field of intelligence, however, predates the 1980s. The military, through ARPA/DARPA, supports the work of American academics, while both the CIA and the Defense Department also monitor global AI research, as evidenced by reports of missions abroad in the 1960s in the context of international colloquia or visits to organizations[29].

Type of action	Precise title	Theme/object	Date	Location
Participation in or organization of symposiums, with contributions from civil experts (industrialists, universities)	*Symposium of Artificial Intelligence Applications to Intelligence Analysis*[30]	AI applications Informing analysts and CIA officials: what is AI, how can it be useful for analysts?	November 30 to December 1, 1982 December 6–8, 1983 (500 people attended) March 19–21, 1985 March 13–14, 1986[31] 1987[32] (*Fifth Intelligence Community Artificial Intelligence Symposium. At the Defense Intelligence Agency, Bolling Air Force Base*)	CIA Headquarters (Washington D.C.); DIA (Bolling Air Force Base) (1985); NBS (Gaithersburg)/ NSA (Fort Meade); etc.

29 United States Intelligence Board, Committee on Documentation, Trip report: CODIB visit to Germany, October 11, 1962, available at: https://www.cia.gov/library/readingroom/docs/CIA-RDP80B01139A000200120018-0.pdf.

30 Science and Technology Division, Office of Scientific and Weapons Research, Directorate for Intelligence, *Symposium of Artificial Intelligence Applications to Intelligence Analysis*, November 18, 1982, available at: https://www.cia.gov/library/readingroom/docs/CIA-RDP85-00024R000500300004-9.pdf.

31 Director of Research and Development, Announcement of AI Symposium, January 29, 1986, available at: https://www.cia.gov/library/readingroom/docs/CIA-RDP89B00297R000400870001-7.pdf.

32 Procurement Management Staff, Weekly report, Period ending on March 31, 1987, available at: https://www.cia.gov/library/readingroom/docs/CIA-RDP89-00063R000300280042-9.pdf.

Type of action	Precise title	Theme/object	Date	Location
Focus groups, coordination	Artificial Intelligence Steering Group (AISG)[33]	Reflections on the potential of AI Regular meetings (June 18, 1984 at NCARAI premises)[34]	1983 (chair: Phil Eckman)	CIA
Focus groups, coordination	Artificial Intelligence Applications Working Group (AIAWG)[35]	Coordinate R&D efforts within the agency to adapt AI to specific intelligence problems	1983	Information Systems Board (CIA)
Relations with American universities	Request for training in artificial intelligence [MAL 87]	Train a member of staff from the agency; represent the intelligence agency in the university institution (integrate students in training; or place "professors" in universities)	1987	Carnegie-Mellon University
Training	Training setup by ISGF	Two- and three-day seminars to introduce AI and its applications in intelligence	1983	Information Science Center (CIA)
	Information technology video seminar. (first use of the headquarters cable TV system in the CIA) [DON 86]	*Seven-part program on AI*	1986	Agency via the headquarters cable TV system

Table 1.3. *Types of actions initiated by the CIA to understand and integrate AI*

33 Created by the IRDC (Intelligence Research and Development Council).

34 NCARAI: Navy Center for Applied Research in Artificial Intelligence.

35 Information Systems Board, Artificial Intelligence applications working group, undated, available at: https://www.cia.gov/library/readingroom/docs/CIA-RDP85-00142R000100030011-8.pdf. Charter of the Artificial Intelligence Applications Working Group of the CIA Information Systems Board, July 1983, available at: https://www.cia.gov/library/readingroom/docs/CIA-RDP85-00142R000100050002-6.pdf.

1.5.2. *Focus groups, research, coordination*

The creation of working, reflection, research and coordination groups, in charge of organizing training initiatives and dialogue with the united industrial world, is neither an approach specific to the CIA, nor to intelligence in general, nor to the theme "AI". In the same year, 1983, the DIA and the NSA, for example[36], also created working groups dedicated to AI. The function of each of these groups was to think about AI for the specific context of each agency (DARPA approaches AI for defense purposes, the CIA for intelligence, etc.). The AISG (Artificial Intelligence Steering Group), within the CIA, observed, in 1984, that defense invests much more in AI research than the intelligence community does, and that a rapprochement should be envisaged in order to capitalize on the advances made by defense[37].

As soon as it was implemented, the ISGF organized meetings which took the form of visits to institutions where AI work was being carried out: the National Bureau of Standards (NBS)[38] was visited on August 27, 1984, and the MITRE Corporation on May 10, 1984[39]. Through these meetings, the members of the ISAG tried to identify projects with potential applications in the field of intelligence.

Within the CIA ISB (Information Systems Board), several work streams were created, including one dedicated to AI (Artificial Intelligence Applications Working Group – AIAWG). During its 1983 meetings, one of the objectives was to identify the CIA's AI needs or the opportunities that AI could offer the agency. The method for identifying these needs was to ask each department within the agency what AI applications they would find useful. This approach assumes that the interlocutors have a fairly precise knowledge of the possibilities of AI: "it was interesting to see common

36 Memo for USDRE from C/AI Steering Group, August 29, 1984, available at: https://www.cia.gov/library/readingroom/docs/CIA-RDP86M00886R000500040004-2.pdf.

37 *Idem*.

38 Artificial Intelligence Steering Group, Meeting Minutes, August 27, 1984, available at: https://www.cia.gov/library/readingroom/docs/CIA-RDP86M00886R000500050002-3.pdf.

39 Artificial Intelligence Steering Group, Trip report, May 10, 1984, available at: https://www.cia.gov/library/readingroom/docs/CIA-RDP86M00886R000500050006-9.pdf.

needs appearing in several offices, but each couched in the language and terminology of their respective application area"[40].

The draft of the summary note of the meeting of December 14, 1983 does not ask any questions about the reasons for this convergence. Perhaps this was due to a real convergence of the needs of the various services consulted. But perhaps it was also the product of a common collective representation of what AI is and what it could provide. The only real differences identified during this exercise remained in vocabulary of the formulations. Furthermore, the requirements were formulated in too general a manner: "Having identified these general requirements, it would appear that we have additional work to do in becoming more specific."[41]

1.5.3. *The network of interlocutors outside the intelligence community*

The integration of AI into the practices and policies of the intelligence community required the involvement of a wide range of external actors.

This approach was neither specific to the new subject of AI nor specific to the intelligence community alone:

> "This country is unique in the ways in which government and private industry work together [...] In the U.S., we rely on a combination of patriotism and profit motive to make our system work [...] New technological devices and new analytical techniques that enable us to understand growing threats to U.S. and its people are based on the synergistic nature of the relations between private industry and government." [DDI 84]

The CIA's approach to AI consisted of keeping abreast of the state of research and development, identifying actors – academic, military and industrial – and engaging in dialog with them in order to understand how available AI could be transferred to the intelligence community. The CIA engaged in dialog with interlocutors outside the intelligence community on a

---

40 AIAWG Meeting Minutes, December 14, 1983, available at: https://www.cia.gov/library/readingroom/docs/CIA-RDP85-00142R000100050004-4.pdf.

41 AIAWG Meeting Minutes, December 14, 1983, available at: https://www.cia.gov/library/readingroom/docs/CIA-RDP85-00142R000100050004-4.pdf.

wide range of topics. The agency remained attentive, ready to respond and alert. For example, in 1983, some 1,200 analysts attended more than 500 conferences on a wide range of subjects [DDI 84].

The CIA was considering adapting knowledge and technology to its needs, and before even thinking about acquiring it (the documents do not specify the methods of transfer envisaged), tried to understand what was at stake:

> "The IASG of the IR&DC is in the process of preparing recommendations for Council consideration concerning where promising Artificial Intelligence technology developments are occurring and which may be of value for adaptation into the Intelligence Community."[42]

This logic guided the entire strategy of acquiring and integrating technology within the CIA, which aimed to save its financial resources and adapt existing technologies to its needs. Thus, faced with requests for acquisition whose justifications are not fundamentally questioned, the director of the CIA wrote, in 1982:

> "Before commenting on the recommendation to acquire a VHSIC/VLSI facility, the Director and I would like the Council to explore a little more exactly what would be involved. Our first blush reaction is prompted by fear that the cost may be too high for the Community to bear. Rather than fund a facility of this nature, possibly we could piggyback on existing commercial facilities to explore circuits which might be unique for intelligence programs […] We believe that the low powered, high density storage technology […] should be pursued by other DoD and commercial programs as well, thus permitting the Intelligence Community to capitalize on developmental work elsewhere."[43]

---

42 Proposed TDY Travel to Boston, October 25, 1985, available at: https://www.cia.gov/library/readingroom/docs/CIA-RDP87M00220R000800930013-8.pdf.

43 IR&DC report, Technology considerations for improved intelligence capabilities, December 1, 1982, available at: https://www.cia.gov/library/readingroom/docs/CIA-RDP87B00305R000801500001-4.pdf.

In 1987, the Deputy Director of Administration of the CIA expressed the desire to place a staff member at Carnegie-Mellon University for a one-year training program to represent the agency technically and politically; the second option proposed was to place a "professor" in residence for two years, with the term in quotation marks:

> "[…] the Agency should have someone at Carnegie-Mellon University who has an ADP background. […] Please […] arrange for an appropriate officer to go to Carnegie-Mellon for a year's training in the field of artificial intelligence or arrange to place an officer at Carnegie-Mellon for two years as a 'professor' in Residence […] It should be an individual who we believe can represent us both technically and politically […] Let's work at this seriously and aggressively." [MAL 87]

	Industry	Universities	Policy/Government
AI Symposium, 1982[44]	Xerox, Symbolics, DEC (Digital Equipment Corporation), VERAC Corporation, System Development Corporation, Artificial Intelligence Corporation, Texas Instruments Corporation, Hughes Research Laboratory	Yale University, University of Pittsburgh, Syracus University	Office of Research and Development (ORD)
Training		Placement of CIA agents as students or professors in American universities	

Table 1.4. *Non-intelligence actors that the CIA solicits and mobilizes and with whom it maintains various types of relationships*

44 Conference program available at: https://www.cia.gov/library/readingroom/docs/CIA-RDP85-00024R000500300005-8.pdf. The program was organized around 14 interventions spread over two days. Four were dedicated to a historical approach to AI research; an introduction and an institutional conclusion; the rest of the days were devoted to technical presentations and industrial demonstrations.

Background	Dates	Vocabulary
Symposium on Artificial Intelligence[45]	1982	Artificial intelligence systems Deductive reasoning Responding to *ad hoc* questions "Smart" data basis Expert systems Target tracking Ocean surveillance Information retrieval
Symposium on Artificial Intelligence	1985	Coping with increasing problems of explosive growth of information
Symposium on AI Applications in the Intelligence Community [ECK 83]	December 8, 1983 [ECK 84] (2nd edition of the symposium, the first was held in 1982, with 300 participants). Symposium sponsored by the ISGF	AI to: - merge, analyze and interpret large volumes of data - increase analyst productivity Natural Language Processing Warning indicators Automated signal processing Expert systems Automated image interpretation

45 The program of this conference is available at: https://www.cia.gov/library/readingroom/docs/CIA-RDP85-00024R000500300005-8.pdf.

Background	Dates	Vocabulary
AIAWG	December 14, 1983[46]	Expert systems Information retrieval, analysis and reporting Networking planning, surveillance and maintenance Consultation, training, analysis and prediction Image interpretation **Natural language** Interface to databases Text retrieval Machine translation Message recognition Message dissemination Speech understanding Speaker recognition Language identification **Image understanding** Optical character recognition Change detection Identifying, counting and classifying objects **Intelligent databases** Inferential reasoning Data resource navigation aids

46 AIAWG Meeting Minutes, December 14, 1983, available at: https://www.cia.gov/library/readingroom/docs/CIA-RDP85-00142R000100050004-4.pdf.

Background	Dates	Vocabulary
AIAWG	December 14, 1983[47]	**Signal interpretation** Real-time TEMPEST surveillance Signal sorting and classification **AI tools and environment** Improvement to information system development process Improved user interfaces Voice input to machines AI system development aids Executive decision aids Administrative system support (budget, travel, personnel)
AI Steering Group[48]	August 29, 1984	Expert systems Natural Language Processing Intelligent database interfaces Image understanding Signal interpretations Geographic and spatial data management Intelligent workstation environments Knowledge acquisition process

47 *Idem.*

48 Memo for USDRE from C/AI Steering Group, August 29, 1984, available at: https://www.cia.gov/library/readingroom/docs/CIA-RDP86M00886R000500040004-2.pdf.

Background	Dates	Vocabulary
Mid-Level Managers' Seminar on Artificial Intelligence[49]	September 11–13, 1984	LISP Semantic networks Production systems Image processing/pattern recognition Image understanding Automating deduction and inference Natural Language Processing Expert systems

Table 1.5. *AI vocabulary within the CIA*

The document is supplemented by a copy of an article published on August 14, 1986, in *The New York Times,* presenting a three-year joint research program between Carnegie-Mellon and IBM. The project was financed by the industrialist to the tune of several million dollars.

The university did not have a candidate imposed on it by the intelligence agency. The Computer Science Department of the university, when asked about the possibilities of hosting a student from the agency, showed itself willing to engage in such an initiative, but outlined conditions: the candidate must have a Master's degree and a prior knowledge of AI. On the other hand, a financial counterpart was requested by the university department [DTE 86].

1.5.4. *What AI applications for what intelligence needs?*

The applications of AI are, for the intelligence community, one of the central points justifying the attention given to these sciences and their developments:

> "The symposium accomplished two things. First, it presented an update on artificial intelligence technologies […] and most

49 *Seminar on Artificial Intelligence*, Information Science Center, available at: https://www.cia.gov/library/readingroom/docs/CIA-RDP89B01330R000600960014-2.pdf.

> importantly, it focused on how these new technologies could be applied within the intelligence community."[50]

The intelligence community is not intended to provide *leadership* in research and focuses on the application dimension. What is in demand is an operational, functional and practical AI.

When the working groups were set up within the CIA in 1982–1983, the first findings stressed that AI was still only at the stage of developing potential applications, not producing operational systems. However, industry efforts could not succeed without a precise definition of intelligence needs. It was therefore a question, at this point, of comparing the potentialities and bringing them to the attention of the intelligence community, who then had to imagine the domains which could benefit from the contributions of AI, then formulate their needs:

> "The contracting parties have started to build teams of qualified staff; they have started to invest in the necessary IT infrastructure; and they are just beginning to understand our specific needs. These efforts will necessarily remain timid until we better understand our own needs."[51]

At the conclusion of one of its meetings in December 1983, the AIAWG produced a list of possible applications of AI in intelligence. This list[52] identified six main areas: expert systems, natural language, image understanding, smart databases, signal interpretation and AI tools and environment.

---

50 *Artificial Intelligence Symposium*, March 28, 1985, available at: https://www.cia.gov/library/readingroom/docs/CIA-RDP87M00539R000600730001-2.pdf.

51 Memo for USDRE from C/AI Steering Group, August 29, 1984, available at: https://www.cia.gov/library/readingroom/docs/CIA-RDP86M00886R000500040004-2.pdf.

52 AIAWG Meeting Minutes, December 14, 1983, available at: https://www.cia.gov/library/readingroom/docs/CIA-RDP85-00142R000100050004-4.pdf.

2

Concepts and Discourses

Constructivist theories give discourse a central place in the analysis of international relations. Discourses construct objects and construct reality. Certain words, as well as certain forms of discourse and particular actors influence and weigh more than others on perceptions, representations and understandings of the world and its issues. International relations, and security and defense issues in particular, are closely dependent on who expresses themselves, when and in what terms:

> "Artificial intelligence is a subject that has fascinated people the world over for many years, but with the introduction of the computer this fascination is becoming a reality. I do not mean the intelligence depicted in so many science fiction books and sensational movies, but rather the slow and methodical application of computers to perform tasks formerly thought to require intelligence." [WIM 77]

The current craze for artificial intelligence is a revival. It can be seen in statements that seek to make an impression: "The development of artificial intelligence could have an effect on society and the economy that is greater than the invention of fire or the industrial revolution."[1]

1 Comments by Enrique Sucar, researcher, quoted in [MAR 18].

The advent of AI is described as a new transformation, a new industrial revolution (it would then be the fourth)[2], a new electricity [NG 16], it would be a Holy Grail to be conquered (we speak of a "quest") [NIL 09], it would bring with it considerable progress (it would be a guarantee of economic growth)[3]. According to Stephen Hawking, it could be the most important development in the history of humankind, and perhaps also the worst. Others formulate the questions differently, questioning its power of creative destruction (destroying the current economic, industrial and production system in order to generate a new one, creating new needs, new jobs and new ways of living in society), or destructive creation (massive destruction of jobs without substitute employment; loss of control over machine-dominated decision-making processes) [MIA 17].

This craze, this excitement, can be observed in the discourse of technology merchants (the industrialists who provide innovations, presented as solutions to urgent challenges or answers to essential needs), as well as in the discourse of politicians, the media who relay information and act as a sounding board, or of "experts". While few voices question the major role that AI will play, dissonances are making themselves heard. Some see it as a positive, albeit sometimes worrying, step forward, while others see it mainly as a fundamental threat. However, this dual point of view is in line with the debates or controversies that may have been sparked by technological developments since the 19th Century, from which developments in computer science in the 1950s and 1960s have not escaped. However, these discourses must be examined for their reflection of the perceptions and appropriation of technologies by societies, and for what they reveal about the way in which technologies and societies are constructed around them.

Some terms, ideas or concepts are more successful than others. But their choices are often decisive, even if the reasons for these choices are sometimes difficult to explain. Ideas, notions and objects could also have been designated or expressed using different terms. This was the case with the term "computer":

> "They should never have been called computers. The first machines were labelled computers because their developers felt

2 The Canadian AI Ecosystem: A 2018 profile, Green Technology Asia Pte Ltd., 2018, available at: http://www.greentechasia.com/wp-content/uploads/2018/02/Canada-AI-Ecosystem-2018-Profile-Summary-Report-Greentech-Asia.pdf.

3 Accenture, "Why Artificial Intelligence is the Future of Growth", June 2016.

> that numerical computations would be their main function […] It might just as well have been called the Oogabooga Box […]. In France, they call all aspects of computer use *l'informatique*, the automatic handling of information. From time to time computer fans have advocated translating the French term and calling computer use 'informatics', an extremely appropriate term. Unfortunately, 'informatics' happens to be the trademark of Informatics.Inc, here in the United States […]. The celebrated scientist John von Neumann got it right at the very beginning, but nobody listened. He called it 'the all-purpose machine'." [NEL 78]

H.A. Simon [SIM 96][4] uses the term "artificial intelligence". Other terms may be used, but are no less problematic:

> [...] you will have to understand me as using 'artificial' in as neutral a sense as possible, as meaning man-made as opposed to natural.
>
> "I shall disclaim responsibility for this particular choice of terms. The phrase 'artificial intelligence' which led me to it, was coined, I think, right on the Charles River, at MIT. Our own research group at Rand and Carnegie Mellon University have preferred phrases like 'complex information processing' and 'simulation of cognitive processes'. But then we run into new terminological difficulties, for the dictionary also says that 'to simulate' means 'to assume or have the mere appearance or form of, without the reality; imitate; counterfeit; pretend'. At any rate, 'artificial intelligence' seems to be here to stay and it may prove easier to cleanse the phrase than to dispense with it. In time it will become sufficiently idiomatic that it will no longer be the target of cheap rhetoric."

2.1. Defining AI

2.1.1. *AI*

Artificial intelligence, like all concepts, is subject to a multitude of definitions. One could rely on and stick to the definitions made by pioneers of

4 This version of the book, dated 1996, is the third. The first one dates from 1976.

the discipline in the 1950s and 1960s. But since then, the concept has never ceased to be questioned, analyzed and deconstructed. Without, of course, ever fully arriving at a single, consensual, ideal definition. Understanding or defining "artificial" on the one hand and "intelligence" on the other hand does not fully define "artificial intelligence".

Defining intelligence was and remains a challenge. It is "an essential function of the mind that allows us to grasp the relationships between things (Latin, *intelligere*; with the prefix *inter*, 'between' and the verb *leggere*, 'to link', 'to put in relation')" [GOF 06].

The term "intelligent/intelligence" declined in the years 1950–1960 in multiple technologies or objects of research. We thus speak of "smart terrestrial satellites" [QUA 57] to distinguish satellites placed in orbit from natural satellites (such as the Moon); intelligent machines, etc., are also discussed.

Marvin Minsky, one of the pioneers of AI, was of the opinion that we should "just use 'intelligence' to mean what people usually mean, namely, the ability to solve hard problems" [MIN 85].

Fred M. Tonge argues:

> "By 'intelligence' we mean a property of a system, a judgment based on observation of the system's behavior and agreed to by 'most reasonable men' as intelligence. 'Artificial intelligence' is then that property as observed in non-living systems. Work directed toward producing such behavior is thereby work in artificial intelligence. While the above is indeed a loose definition, more useful in suggestiveness than in precision, it should suffice for our purposes. It does contain at least one strong assumption – that, *a priori*, intelligence is not restricted to 'living' systems. And it does suggest that, if the issue of whether artificial intelligence does or will ever (or can ever) exist is really worthy of further argument or concern, then some agreement should be reached concerning the rules for observation and agreement among most reasonable men. Here, however, we shall treat the term as merely a convenient label." [TON 66]

The lack of consensus on the definitions of the two concepts of intelligence, on the one hand, and artificiality, on the other hand, makes the task of defining their combination in the expression “artificial intelligence” more difficult.

Table 2.1 provides a long list of definitions, in chronological order, starting in the early 1960s.

This already lengthy but non-exhaustive compilation of AI definitions calls for a few remarks. According to these definitions, artificial intelligence is sometimes a science or branch of the computing discipline; sometimes a family of technologies; sometimes the ability of machines to think like humans. But if we retain the definition posed by Mr. Minsky in 1960, artificial intelligence does not exist today, because no machine, no algorithm, no computer is yet capable of critical thinking. The capacities of modern computers consist in massively, rapidly calculating, solving equations, learning, “reasoning”, but all this is not enough to constitute a capacity for critical thinking.

Definition	Source	Year
Artificial intelligence is “[...] the science of making machines do things that would require intelligence if done by men. It requires high-level mental processes such as: perceptual learning, memory and critical thinking.”	[MIN 61]	1960
“Aptitude of a machine to give an answer to a question that is usually obtained with (human) reasoning guided by memory, experience and intuition.”	[RAY 63]	1963
“Artificial Intelligence is the experimental and theoretical study of perceptual and Intellectual processes using computers. Its ultimate goal is to understand these processes well enough to make a computer perceive, understand and act in ways now only possible for humans.”	[MCC 71]	1971
“The branch of computer science called Artificial Intelligence involves the attempt to program or build a digital computer capable of producing behavior that a human being would accept as truly intelligent.”	[SAL 76]	1975
“Artificial intelligence [is] the efforts to emulate and bolster human reasoning processes as programs in computers.”	[LED 76]	1976

Definition	Source	Year
"Artificial Intelligence is not the study of computers, but of intelligence in thought and action. Computers are its tools, because its theories are expressed as computer programs that enable machines to do things that would require intelligence if done by people."	[BOD 77]	1977
"Artificial intelligence (hereinafter AI) can be defined as the capability of a device to perform functions that are normally associated with human intelligence, such as reasoning, learning and self-improvement."	[WIL 83]	1983
"Artificial intelligence (AI) is the field of scientific inquiry concerned with designing machine systems that can simulate human mental processes."	[LAW 84]	1984
"An interdisciplinary subfield of computer science that seeks to recreate in computer soft-ware the processes by which humans solve problems."	[AND 84]	1984
"Artificial Intelligence technology is intended to design and to produce intelligent computer systems, that is, computers that can imitate the human thought process, yet attain more accurate results."	[DAN 86]	1986
"Artificial intelligence will be considered the branch of computer science dealing with the development of methods that solve problems previously considered solvable only by humans. These problems include reasoning, understanding, natural language, planning and problem solving, learning, visual processing and manipulation of the physical world."	[FRA 86]	1986
"AI is a family of technologies that use stored expertise and knowledge. With these technologies, knowledge, a vital corporate resource, can be captured and tested, improved and replicated, used and reused."	[MEL 89]	1989
"AI is the computer-based exploration of methods for solving challenging tasks that have traditionally depended on people for solution. Such tasks include complex logical inference, diagnosis, visual recognition, comprehension of natural language, game playing, explanation, and planning."	[HOR 90]	1990

Definition	Source	Year
"I call artificial intelligence, the computer technology that aims to simulate the intelligent behavior of human beings, i.e. the behavior that enables man to solve the problems facing him, whether intellectual or pragmatic, either systematically and consciously or intuitively and unconsciously. Artificial intelligence is the heir to several traditions, the main ones being biology, psychology, computer science and linguistics."	[JOR 94]	1994
"Artificial intelligence means devices, software and systems that are able to learn and make decisions in almost the same manner as humans. Artificial intelligence allows machines, devices, software, systems and services to function in a sensible way according to the task and situation at hand."	Artificial Intelligence Program (Finland) [AIF 19]	2017
"Artificial intelligence is the science of building computer programs that aim to perform tasks that would require some intelligence if they were done by human beings."	[VIL 18]	2018
"AI is a constellation of technologies that enable machines to act with higher levels of intelligence and emulate the human capabilities of sense, comprehend and act."	National Strategy for Artificial Intelligence (India)[5]	2018
"AI is a combination of technologies that allow smart machines to support human capabilities and intelligence by sensing, comprehending, acting and learning; thereby allowing people to achieve much more than can be achieved without AI."	Accenture's definition, cited in the Mauritius National Strategy[6]	2018
"The branch of computer science focused on programming machines to perform tasks that replicate or augment aspects of human cognition."	Director of National Intelligence (United States)[7]	2018
"AI consists of a computer program that is capable of executing decisions in a faster, more intelligent manner than a human based on a defined set of parameters and a clear objective."	[EAT 18]	2018

5 NITI, National Strategy for Artificial Intelligence, June 2018, available at: https://niti.gov.in/writereaddata/files/document_publication/NationalStrategy-for-AI-Discussion-Paper.pdf.

6 Working Group on artificial intelligence, Mauritius Artificial Intelligence Strategy, November 2018, available at: http://mtci.govmu.org/English/Documents/2018/Launching%20Digital%20Transformation%20Strategy%20191218/Mauritius%20AI%20Strategy%20(7).pdf.

7 Director of National Intelligence, The AIM Initiative, 2018, available at: https://www.dni.gov/files/ODNI/documents/AIM-Strategy.pdf.

Definition	Source	Year
"Artificial intelligence is systems based on algorithms (mathematical formulae) that, by analyzing and identifying patterns in data, can identify the most appropriate solution."	National Strategy for Artificial Intelligence (Denmark)[8]	2019
"Artificial intelligence (AI) is a diverse field of computer science that is constantly evolving [...]" "Artificial intelligence (AI) refers to systems that display intelligent behavior by analyzing their environment and taking actions – with some degree of autonomy – to achieve specific goals."	National Artificial Intelligence Strategy of the Czech Republic[9] The second part of the definition, relating to the Lithuanian strategy, is that of the European Commission	2019
"In this section, the term 'artificial intelligence' includes the following: (1) Any artificial system that performs tasks under varying and unpredictable circumstances without significant human surveillance, or that can learn from experience and improve performance when exposed to data sets. (2) An artificial system developed in computer software, physical hardware, or other context that solves tasks requiring human-like perception, cognition, planning, learning, communication, or physical action. (3) An artificial system designed to think or act like a human, including cognitive architectures and neural networks. (4) A set of techniques, including machine learning, that is designed to approximate a cognitive task. (5) An artificial system designed to act rationally, including an intelligent software agent or embodied robot that achieves goals using perception, planning, reasoning, learning, communicating, decision making, and acting."	John S. McCain National Defense Authorization Act for Fiscal Year 2019[10]	2019

Table 2.1. *Definitions of AI by year and chronological order*

8 The Danish Government, National Strategy for Artificial Intelligence, Ministry of Finance and Ministry of Industry, Business and Financial Affairs, March 2019, available at: https://eng.em.dk/media/13081/305755-gb-version_4k.pdf.

9 National Artificial Intelligence Strategy of the Czech Republic, Ministry of Industry and Trade of the Czech Republic, 2019, available at: https://www.mpo.cz/assets/en/guidepost/for-the-media/press-releases/2019/5/NAIS_eng_web.pdf.

10 https://www.govinfo.gov/content/pkg/BILLS-115hr5515enr/pdf/BILLS-115hr5515enr.pdf.

The term "machine", very often used in these definitions, will no doubt also have contributed to being misleading. For what exactly is meant by "machines"? For mathematicians and computer scientists, the machine can be an abstract model (as is the Turing machine). But more generally, the machine is a "manufactured complex object capable of transforming one form of energy into another and/or of using this transformation to produce a given effect, to act directly on the object of work in order to modify it according to a fixed goal"[11]. But the machine is also synonymous with a calculator, a computer. We still speak of machine language or a machine program in computer science. Which of these meanings is a precise reference to each use of the term? The allusion to the object transforming energy or operating with energy can just as easily refer to electrical tools or instruments, computers, as it can robots.

The primary purpose of these machines, as F.H. Raymond for example expresses it, is to give answers to questions. The calculating machine is a machine that provides answers (the results of calculations) to questions (which are operations). Of course, we can imagine that the intelligent computer goes further, beyond the strict framework of mathematical calculation. However, in today's predictive tools, what else is there to do except perform calculations on large masses of data? The answers can also be called "solutions", and the questions "problems". Artificial intelligence should broaden the range of questions or problems that can be submitted to machines. This is the goal that expert systems would pursue in the 1980–1990s with varying degrees of success. But the principle of answering questions remains today at the heart of the aims assigned to AI. Predicting means answering the questions "who will… ", "what will… ", "how will…" (the weather, the economy, market trends, crowd behavior, etc.). Detecting information with the help of so-called intelligent cameras means answering the questions "where will…" (an individual, an object, the signs of disease on an ultrasound, etc.), "who is there… ", "what can you see… ?". But human intelligence means more than just solving problems or providing answers to questions.

Human intelligence is the only form of intelligence targeted by AI, the one to be imitated, reproduced, attained, the one that serves as a reference, from which humans can estimate whether the machine is intelligent or not. Artificial intelligence aims to reproduce in the computer, through software, human

11 Definition produced by the LTRC: https://www.cnrtl.fr/definition/machine.

capacities or functions (understanding, reasoning, learning, perception, decision-making, etc.). The possibility of taking animal intelligence into consideration or as a model never seems to be considered. The human remains the model, the reference point, which must be copied and attained, and we rely on human intelligence to judge that of the machine. For some years now, the idea has been circulating that AI has overtaken humans or is about to do so, thus making it useless, pushing humans out of the loop, creating a field in which the human has no place because they are unable to compete with the capacities of machines that stir up volumes of data at considerable speeds, out of reach of the capacities of the human brain, as well as because the machine would proceed with reasoning that is now alien to a human. Humans, who were a model, an objective, a reference for AI since the 1960s, would thus be obsolete and would no longer be this typical model. Humans would no longer be the pivot from which intelligences and abilities are assessed.

Figure 2.1. *Cloud of terms built up from the set of definitions*[12] *in Table 2.1. For a color version of this figure, see www.iste.co.uk/ventre/artificial.zip*

2.1.2. *Expert systems*

Expert systems are one of the key concepts in the history of artificial intelligence. We mention them many times in this book. It is worth highlighting a few definitions.

12 The cloud is made from the English language definitions, which are the core of this compilation.

The terms on which expert system definitions are based differ somewhat from those used for definitions of artificial intelligence. The dominant terms are "human", "knowledge", "problem", "reasoning" and "solution". Only the concept "human" is common to the two most common sets of terms. Humans, with their expertise, interact with the machine. They are indispensable to the system as a whole. Together, human and machine must find solutions to complex, difficult problems. Knowledge (the basis of knowledge) is the key element of these systems.

The applications of expert systems have been numerous since the 1970s: in 1972, for example, in the field of medicine, the MYCIN system was developed to identify the organisms responsible for an infection based on patients' symptoms (Stanford University); in the military field, expert systems have provided decision support tools, etc.

Definition	Source
"An 'expert system' is an intelligent computer program that uses knowledge and inference procedures to solve problems that are difficult enough to require significant human expertise for their solution."	[FEI 80]
"Expert systems is a relatively new sub-discipline of artificial intelligence and has to do with using computers to imitate the reasoning or judgment process of human experts."	[BAC 87]
"A typical expert system consists of three components. First, a 'knowledge base', containing information about a highly specific field of expertise. Second, an 'inference engine', which must make decisions about which parts of that knowledge base are relevant to solving a given problem and then piece together a line of reasoning linking the problem with its possible solution. Finally, the third component of an expert system that determines its role as a consultant is the 'user interface', which allows human experts to interact with the machine and to request it to explain the rationale for its different choice."	[DEL 91]
"One of the most successful branches produced by Artificial Intelligence research. In these systems the ability to reason in a logical way characteristic of all AI programs is complemented by the problem-solving abilities of human experts in particular fields. Knowledge banks are then provided with a human interface to allow them to play the role of 'mechanical advisers': given a problem in a very specific field, these systems can give expert advice regarding a possible solution, and even provide their human users with the line of reasoning followed to reach a particular piece of advice."	[DEL 91]
"A computer program capable of considering a vast body of knowledge, reasoning, and then recommending a course of action."	[HAN 92]

Table 2.2. *Definitions of expert systems*

Despite their success, expert systems have been the subject of much criticism, in particular because of the mistakes they can make if the problems raised go beyond the strict framework of their area of expertise. For example, when faced with a victim of a gunshot wound, the symptom is the bleeding, the expert system specializes in searching for infectious organisms and it will search for bacteria responsible for the bleeding. The expert system is unable to leave its field on its own. The expert system only works in its own little world. It also has no common sense, it is unable to understand or analyze the effects of its recommendations on the context.

Figure 2.2. *Cloud of terms built from the expert system definitions in Table 2.2. For a color version of this figure, see www.iste.co.uk/ventre/artificial.zip*

2.1.3. *Machine learning and deep learning*

Machine learning or automatic learning was a term introduced in 1959 by Arthur Samuel. Machine learning is a sub-domain of artificial intelligence. The research community has several specialized journals, such as the *Machine Learning Revue* (published by Springer, founded in 1986) and the *Journal of Machine Learning Research* (founded in 2000)[13]. The concept

13 The articles, starting from volume 1, are available on the journal's website: http://jmlr.csail.mit.edu/papers/.

covers the idea that software or a machine is able to learn without everything being explicitly programmed into it in advance. From data, from particular observations, the algorithm looks for a link or relationships in this data. Learning only a first phase, which then makes it possible, by relying on the acquired knowledge, to recognize, from "predicting", to extrapolating. Facebook or Google, for example, use such algorithms to try to define and propose connections, to identify the information that most interests the Internet user. The algorithm thus sorts through the masses of data and information in place of the Internet user. Credit card companies can use machine learning to detect fraud. Machine learning is used for object recognition in images (facial recognition, scene analysis, etc.). "Deep learning is a subcategory of machine learning algorithms that use multi-layered neural networks to learn complex relationships between inputs and outputs" [SHA 18a].

Deep learning is one of the implementation techniques of machine learning, which relies more specifically on neural networks:

> "Deep learning is a type of machine learning that can process a wider range of data resources, requires less data preprocessing by humans, and can often produce more accurate results than traditional machine learning approaches."[14]

2.1.4. *The robot, robotics*

AI and robotics are closely linked both in research and in our imagination, notably because of the literary production of science fiction which, since the beginning of the 20th Century, has never ceased to offer us universes populated by robots, endowed with an intelligence equal, if not superior, to that of humans, and always presenting a challenge for the latter. From the few definitions retained (Table 2.3), we will say the robot:

– is mainly a machine;

– is programmable;

– has a specific function;

14 https://niti.gov.in/writereaddata/files/document_publication/NationalStrategy-for-AI-Discussion-Paper.pdf.

– has a degree of freedom that is controlled and decided by the humans (who program it);

– has a resemblance to humans which is related to its functionality (seeing, moving, grasping objects, etc.), more than its appearance.

Definition	Source	Year
"A manipulating industrial robot is an automatically controlled, reprogrammable, multi-purpose, manipulative machine with several degrees of freedom, which may be either fixed in place or mobile for use in industrial automation application."	International Organization for Standardization, ISO	
"Automatic apparatus capable of manipulating objects or performing operations according to a fixed, modifiable or adaptable program."	Larousse Dictionary (online)	
"In works of fiction, [it is] a machine, an automaton with a human aspect capable of acting and speaking like a human being."	CNRTL (online)	
"A machine carrying out, thanks to a microprocessor-based automatic control system, a precise task for which it was designed in the industrial, scientific or domestic field."	CNRTL (online)	
"A machine that resembles a living creature in being capable of moving independently (as by walking or rolling on wheels) and performing complex actions (such as grasping and moving objects)."	Merriam-Webster (online)	
"A device that automatically performs complicated, often repetitive tasks (as in an industrial assembly line)."	Merriam-Webster (online)	
"A robot is a reprogrammable multifunctional manipulator designed to move materials, parts, tools or specialized devices through variable programmed motions for the performance of a variety of tasks."	[SCI 00]	2000
"Robots are sensomotoric machines for the extension of human mobility. They consist of mechatronic components, sensors and computer-based control functions."	[CHR 01]	2001
"A simple definition of a robot is: 'any machine programmed to do work'."	[ELL 12]	2012

Table 2.3. *Definitions of "robot"*

Terms	Definition
Android	Has the appearance of a human (male). More generally, a robot that has the appearance of a human being. Term used since the 19th Century (used in *The Future Eve* by Auguste Villiers de l'Isle-Adam, in 1886; in *The Brazen Android*, by William Douglas O'Connor, 1891).
Ginoid	Looks like a woman. Example: Maria, the robot in the movie *Metropolis.*
Humanoid	Looks like a human (morphologically).
Cyborg	Cybernetic organism. A living being to which artificial, electronic or mechanical parts have been grafted or implanted. It differs from the robot, which is entirely artificial and mechanical.
Droid	The term comes from science fiction, which designates robots with a shape reminiscent of androids.
Cobot	Collaborative robot (robot–human interaction).
Chatbot	Agent that dialogues with a user.

Table 2.4. *Some terms related to the "robot" world*

Several ideas are thus associated with the "robot":

– It is designed to help, to assist humans: the automatons of the 17th and 19th Centuries are machines similar to humans and have an entertainment function, on the one hand (by playing an instrument, or by the quasi-magical nature of their existence and the technical prowess of their creators), and, on the other hand, allow people to imagine a future in which the machine will carry out tasks for people.

– It is the symbol of a future world: the robot is present in futuristic stories where terrestrial and extraterrestrial universes are mixed. It is omnipresent in spatial conquest.

– It is an existential threat: Karel Capek's robot (1920) is the archetype of the robot that constitutes a threat to humanity because it will enslave it. The robot first frees humans from their tasks, then the roles are reversed; the humans' creation is turned against them. The robot becomes monstrous.

– It is an actor of conflict, of violence: the science-fiction news robot is often a fighter who confronts humans or other robots. It confronts other heroes: in 1963, Russ Manning created an anti-robot hero, Magnus, Robot Fighter; also read *The Human Bat V the Robot Gangster* (Edward R. Hime-Gall).

– It is a pretext for reflections on the human condition, on the relationship between human and machine: if people have their own laws, rules, principles, morals and ethics, can't robots have them too?

2.2. Types of AI

Typologies are used to classify, order and create categories, and provide reading grids that allow a concept to be grasped in its multiple facets. The exercise is necessarily simplifying and can be marked by subjectivity.

One of the most common typologies, devised in the 1960s, is based on the distinction between:

– weak or narrow AIs, which are AIs dedicated to a specific task or function. These machines do not learn by themselves, all rules must be given to them beforehand (example: all the rules of a game). The weak AI is the one that can be produced in the current state of knowledge and can therefore be the subject of State AI strategies: "The Federal Government has oriented its strategy to the use of AI to solve specific problems, i.e. to the 'weak' approach"[15];

– strong or broad/general AI, able to learn by itself and then decide and act as an individual would. Such AIs do not yet exist, ones that would be able to think in a general way, and make decisions based on learning and not training. Several tests define general AI: the Turing test; the coffee test formulated by Steve Wozniak at 2007; the Robot University Student Test; and the employment test (formulated by Nils John Nilsson in 2005).

Some authors add a third level, proposing: weak AI, general artificial intelligence and super artificial intelligence or superior artificial intelligence:

> "Superintelligence is AI that is much smarter than humans. Superintelligence does not currently exist, but it has been proposed that it could someday be built, with massive and potentially catastrophic consequences." [BAU 18]

15 Artificial Intelligence Strategy, The Federal Government, November 2018, available at: https://www.ki-strategie-deutschland.de/home.html?file=files/downloads/Nationale_KI-Strategie_engl.pdf.

This level of AI is higher than general AI: the general AI would have an intelligence equal to that of a human; superintelligence would exceed this. Superintelligence gives rise to essentially catastrophist discourses, on the theme of existential threat, expressing the fear of an overtaking, an overflow and a total loss of control by humans over an intelligence whose actions could then no longer be anticipated or controlled by anyone.

	High AI (strong, general) or AGI (Artificial General Intelligence)	**Low AI (weak, narrow) or ANI (Artificial Narrow Intelligence)**
Definition	Refers to any autonomous AI system that has intellectual capacities equal to or greater than those of a human being.	The AI system is focused on solving specific problems. These machines operate within the strict framework of the scenarios they have been programmed for.
Intelligent machines produced	None.	AlphaGo (Google), Watson (IMB), Siri (Apple), etc. Most of the AIs produced to date fall into this category of low AIs.

Table 2.5. *Typology of AI*

Another form of two-branch typology distinguishes AI type 1 and type 2:

– type 1 groups the two types of AI mentioned above (low and high AI);

– type 2 is different:

- reactive machines, which do not retain the memory of the past to alter their decisions or future choices,

- those with limited memory, which use information from the past to inform future decisions (found, for example, in autonomous vehicle driving systems, where environmental observations are not permanently stored),

- AI which can "understand" the emotions, behaviors, thoughts of individuals, and able to interact socially,

- finally, those that would have self-awareness, in the image of the human being.

A distinction is therefore made between AI that is feasible in the current state of knowledge and AI that is not feasible in the short and medium terms.

The typologies may consist of breaking down the fields of application for AI, or of its main areas of research and development: "Artificial intelligence contains the following major subfields: expert or knowledge based systems, natural language processing, artificial or computer vision, sound sensing/ understanding and robotics." [ALB 88]

2.3. Evolution of the themes over time

2.3.1. *Google Trends*

An initial curve drawn from Google Trends data on worldwide Internet user queries shows an initial downward phase over the period 2004–2019, from 2004 to around 2007–2008, which remains at a low level until 2014–2015, at which point the trend moves upward. However, this last phase does not appear particularly dazzling on this curve. The renewed interest appears to be gradual, and has not yet reached the 2004 level.

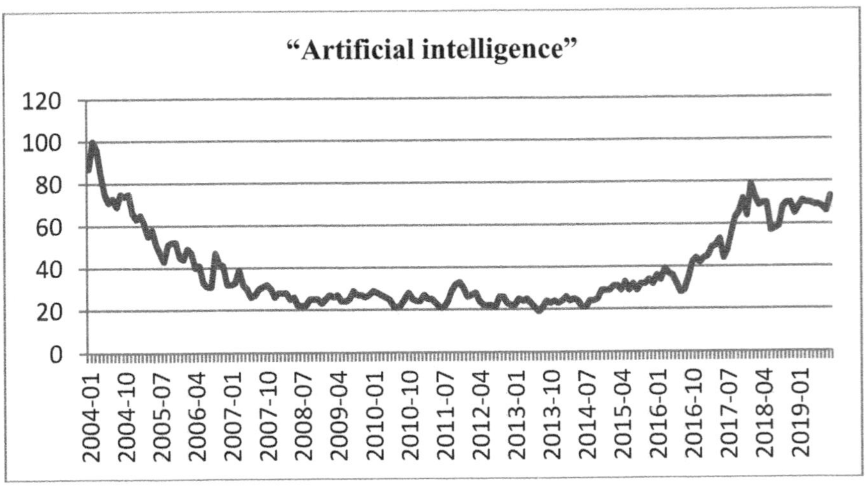

Figure 2.3. *Google Trends. Evolution of queries in the world related to "artificial intelligence" (data recorded September 25, 2019)*

2.3.2. *The AAAI magazine*

AAAI, the American Association for Artificial Intelligence, was founded in 1979. It was later renamed the Association for the Advancement of Artificial Intelligence. Since 1980, the association has published *AI Magazine*, all issues of which are available on its website[16]. We have searched throughout this corpus (1980–2019) for some terms related to the field of artificial intelligence, in order to observe the evolution of their use by the research community. The observation we made only concerns the article titles and therefore does not include either the abstracts or the contents of the articles in their entirety. Perhaps we would obtain significantly different results by analyzing the full contents. But observing the titles of the articles, reflecting the central theme of the research presented, seems to us to be sufficiently relevant at this stage[17].

We thus note that the concept of an "expert system" appears in the period 1980–2000 and then no longer appears in the article titles. The subject seems to have fallen into disuse.

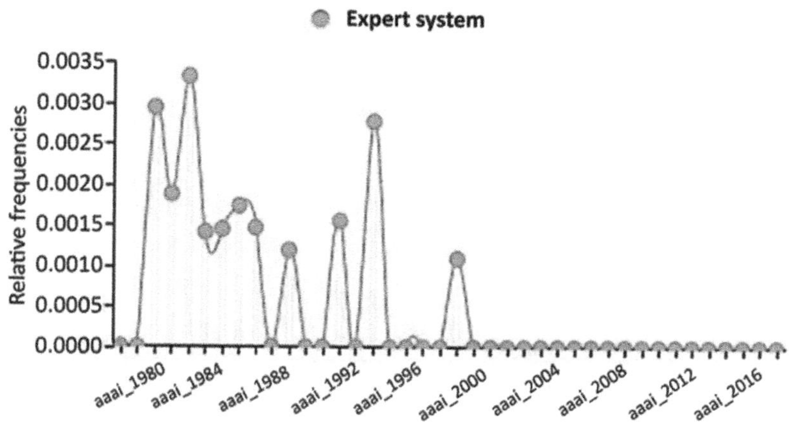

Figure 2.4. *Evolution of the presence of the concept of "expert systems" in AAAI publications (1980–2019)*

The theme of "Machine Learning" is more stable and constant over time (Figure 2.5).

16 https://aaai.org/ojs/index.php/aimagazine/issue/archive/.

17 Data processing was carried out using the online analysis tool Voyant Tools.

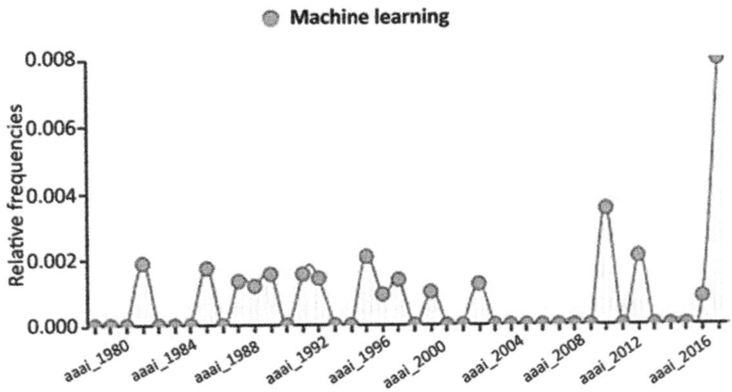

Figure 2.5. *Evolution of the presence of the "Machine Learning" concept in AAAI publications (1980–2019)*

The topic of robotics (robot, robotics) is also treated in a relatively continuous manner. There is therefore no particular phase of enthusiasm, sudden interest or, on the contrary, abandonment of these subjects (Machine Learning, robotics), as opposed to the phenomenon that seems to affect the treatment of expert systems (Figure 2.6).

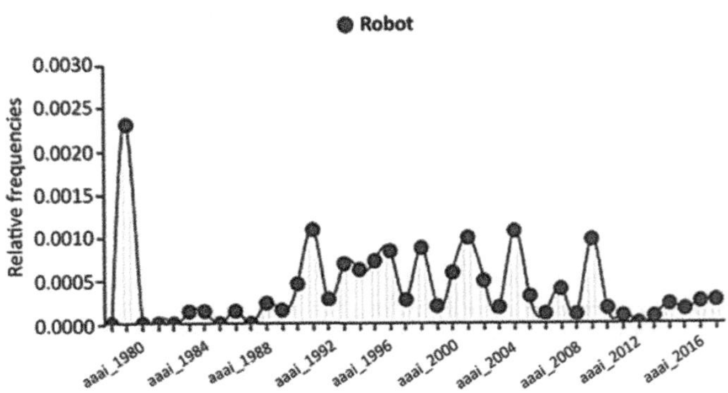

Figure 2.6. *Evolution of the presence of the "robot" concept in AAAI publications (1980–2019)*

The theme of autonomy came to the fore in the early 1990s (Figure 2.7).

Then, there are subjects or themes that seem to be totally absent from published research. This is the case with the terms "war", "warfare", "army", "UAV", "drone", "hack", "malware", "justice", "power", etc.

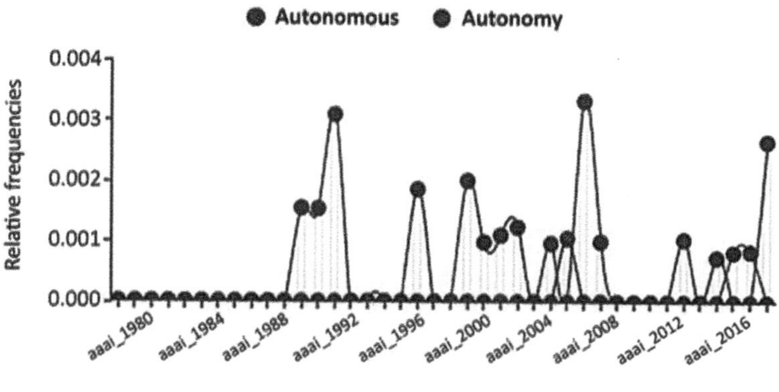

Figure 2.7. *Evolution of the presence of the "autonomous/autonomy" concept in AAAI publications (1980–2019). For a color version of this figure, see www.iste.co.uk/ventre/artificial.zip*

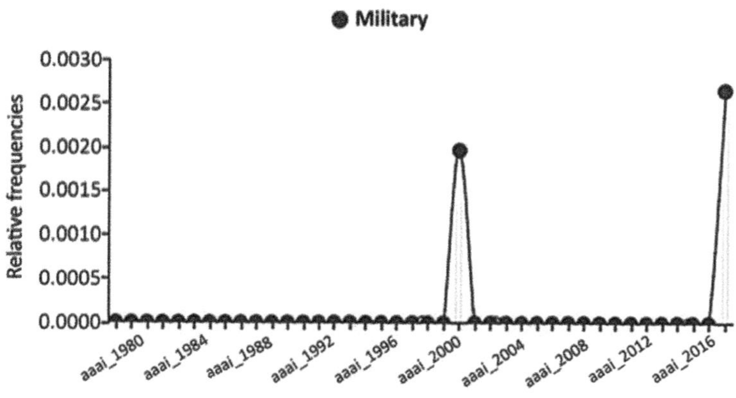

Figure 2.8. *Evolution of the presence of the "military" concept in AAAI publications (1980–2019)*

The military field occupies little space in articles published by the association's journal. The term "military" is only mentioned in two articles:

– the first, in 2002, was entitled "Interchanging agents and humans in military simulation";

– the second, in 2019, was entitled "Artificial intelligence, robotics, ethics, and the military: A Canadian perspective".

The term "cyber" did not appear until 2019 ("Artificial intelligence and game theory models for defending critical networks with cyber deception").

The issue of security was only recently introduced and is still the subject of only a few articles (Figure 2.9).

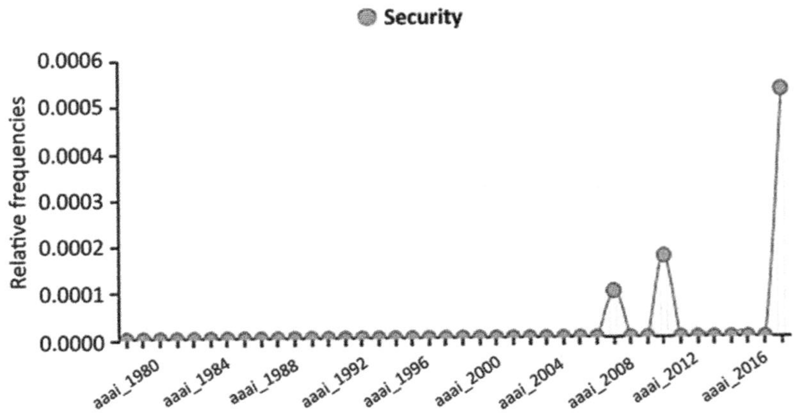

Figure 2.9. *Evolution of the presence of the concept of "security" in AAAI publications (1980–2019)*

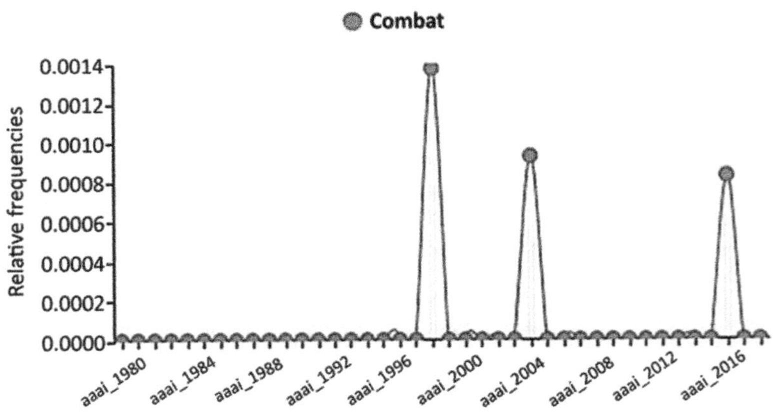

Figure 2.10. *Evolution of the presence of the concept of "combat" in AAAI publications (1980–2019)*

The military institution is also little mentioned in the relevant subjects:

– the U.S. Air Force was discussed in a single article in 1985 ("Artificial intelligence research capabilities of the Air Force Institute of Technology"); and the Navy was discussed in a 1991 article on "The use of Artificial intelligence by the United States Navy: Case study of a failure";

– the dimension of combat is taken into account in the work on simulation tools:

- "Automated intelligent pilots for combat flight simulation" (1999),

- "Synthetic adversaries for urban combat training" (2005),

- "PAWS – A deployed gametheoretic application to combat poaching" (2017).

Legal and ethical considerations were integrated into the work of the AI community during the 1980s, abandoned for a time and then taken up again by the AI community in 2005.

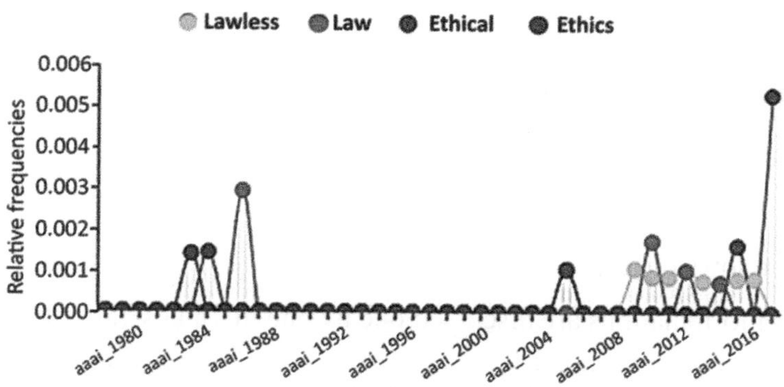

Figure 2.11. *Evolution of the presence of the concepts "law" and "ethics" in AAAI publications (1980–2019). For a color version of this figure, see www.iste.co.uk/ventre/artificial.zip*

2.4. The stories generated by artificial intelligence

Artificial intelligence, like the computer or more broadly computer science, has given rise to passionate debates, sometimes controversies, philosophical, political, sociological, economic, historical, ethical and legal reflections. We cannot list all the questions raised by AI in an exhaustive manner, but we will nevertheless mention some of them.

2.4.1. *The transformative power of AI*

> "Weizenbaum doesn't name any specific task that computers cannot carry out, because he wishes 'to avoid the unnecessary,

interminable, and ultimately sterile exercise of making a catalogue of what computers will and will not be able to do, either here and now or ever'." [MCC 76]

An inventory of the sometimes clear-cut opinions on what artificial intelligence can or cannot do, what it transforms, what it impacts, seems useful to us, however. For it enables us to understand how science and technology take their place in society and around which arguments the debates are built. Here, we attempt to list the opinions formulated about AI, identified in very diverse contents (press articles, blogs, technical journals, etc.). We have grouped these discourses around the following forms of expression:

– AI is.../AI is not...;

– AI will.../AI will not...;

– AI should.../should not...;

– AI produces/makes...;

– AI allows/enables...

These formulations are discussed one by one in Tables 2.6–2.11.

Artificial intelligence is...	Date
"Artificial Intelligence (AI) is transforming the nature of almost everything which is connected to human life e.g. employment, economy, communication, warfare, privacy, security, ethics, healthcare etc."[18]	2016
"AI is the practice that intends to make machines think."[19]	2017
"Artificial intelligence is likely to be a 'big help' rather than a 'hindrance'."[20]	2018
"Artificial intelligence is changing every aspect of war."[21]	2019
"How artificial intelligence is changing cyber security landscape and preventing cyber attacks."[22]	2019

Table 2.6. *Some assertions or assumptions about AI intakes*

18 https://papers.ssrn.com/sol3/papers.cfm?abstract_id=2836438.

19 https://www.youtube.com/watch?v=RpZEqbZQUcA.

20 https://www.cnbc.com/2018/11/14/artificial-intelligence-will-not-replace-people-salesforceexecutive.html.

21 https://www.economist.com/science-and-technology/2019/09/07/artificial-intelligence-is-changing-every-aspect-of-war.

22 https://www.entrepreneur.com/article/339509.

Artificial intelligence is not…	**Date**
"Artificial intelligence is not supposed to be a carbon copy of human life."[23]	Not dated
"Artificial intelligence is not a distinct technology. It depends for its power on a number of prerequisites: computing power, bandwidth, and large-scale data sets, all of which are elements of 'big data', the potential of which will only be realized using artificial intelligence. If data is the fuel, artificial intelligence is the engine of the digital revolution."[24]	2015
"Artificial Intelligence is not new, it was first coined by the American scientist John McCarthy in 1955, who is also considered co-founder of the field Artificial Intelligence." [25]	2016
"Artificial intelligence is not a threat to human society."[26]	2017
"Artificial Intelligence is not an existential threat."[27]	2017
"Artificial Intelligence is not about machines ruling our humans, but machines and humans working together."[28]	2017
"Artificial Intelligence is not the future, it is already happening and widely available."[29]	2017
"Artificial intelligence is not the same as artificial consciousness (artificial consciousness has sometimes been called 'Strong AI' or 'Full AI')."[30]	2017
"Artificial Intelligence is not neutral: there are cases where software used in justice systems have been racially biased."[31]	2018
"Artificial intelligence is not just for robots."[32]	2018
"Artificial intelligence is not the end of all jobs."[33]	2018

23 https://www.hypergiant.com/capabilities-of-artificial-intelligence/.

24 https://assets.publishing.service.gov.uk/government/uploads/system/uploads/attachment_data/file/566075/gs-16-19-artificial-intelligence-ai-report.pdf.

25 https://vinodsblog.com/2018/03/11/the-exciting-evolution-of-machine-learning/?fbclid=IwAR3WUFDuTaOP78HYlFVhvAVqYe4kS2D_O37p48VIs3M11LKrqJqO8m6jm0A.

26 https://www.reasoning.world/artificial-intelligence-is-not-a-threat-to-society/.

27 Shermer M., "Why artificial intelligence is not an existential threat", *Skeptic Magazine*, vol. 22, no. 2, pp. 29–35, 2017, available at: https://pocketmags.com/skeptic-magazine/222/articles/144321/why-artificial-intelligence-isnot-an-existential-threat.

28 http://www.stantive.com/Artificial-Intelligence-Empowering-People-Not-The-Rise-Of-The-Machines.

29 https://www.pwc.at/de/publikationen/verschiedenes/artificial-intelligence-in-hr-a-no-brainer.pdf.

30 http://www.pcts.org/meetings/2017/PCTS2017Nov-Green-ReflectionsAI.pdf.

31 https://www.youtube.com/watch?v=zDHSD4twyIA.

32 http://yp.ieee.org/artificial-intelligence-is-not-just-for-robots/.

33 https://www.population.sg/articles/artificial-intelligence-is-not-the-end-of-all-jobs.

Artificial intelligence is not...	Date
"What we call Artificial Intelligence is not real intelligence. Only human intelligence is the real thing."[34]	2018
"Artificial intelligence is not a silver bullet for cyber security."[35]	2018
"Artificial intelligence is not so intelligent."[36]	2018
"Artificial Intelligence is not the best way to curtail online hate."[37]	2018
"Artificial intelligence is not necessarily a straightforward topic for governments across the continent."[38]	2018
"Artificial Intelligence is not new, but the availability of computing power, new technologies to capture enormous amounts of data points and the speed by which we can connect, store and analyze data give new insights and opportunities that can lead to business performance breakthroughs, new business models and changing markets. Thanks to the combination of the technologies that are available today and the speed by which the industry can learn from its processes, products and customers will increase exponentially."[39]	2018

Table 2.7. *Some assertions or hypotheses about what artificial intelligence is not*

The rhetoric surrounding the notion of AI emphasizes the principle of social change that the development and integration of these technologies into a wide range of areas of social activity involves or will involve. These changes are considered to be of varying intensity, sometimes referred to as "changes", "transformations", "revolutions" and "breakthroughs".

Asserting that AI will (or will not) do, that it will (or will not) transform, is deterministic (technological determinism holds that technology determines social change). The relationship between technology and the individual or society can also be viewed in the opposite direction, that of social determinism, i.e. considering the social conditions that allow technology to

34 https://www.morningfuture.com/en/article/2018/06/04/emanuele-severino-roger-penrose-artificial-intelligence-natural-intell/331/.

35 https://www.computing.co.uk/ctg/news/3037214/artificial-intelligence-is-not-a-silver-bullet-for-cyber-security.

36 https://medium.com/forcit/artificial-intelligence-is-not-so-intelligent-according-to-steve-wozniak-b4c26100bfdc.

37 https://globalnews.ca/news/4461472/using-artificial-intelligence-to-fight-online-hate/.

38 https://qz.com/africa/1305211/google-is-making-a-big-bet-on-artificial-intelligence-in-africa-with-its-first-research-center/.

39 http://www.euromanuforum.com/documents/EFM_Papers_Presented_200618.pdf.

be sold, developed, used (and in particular in what forms). It is then society that produces the technology. And since society cannot be reduced to a homogeneous block, technology will produce different effects from one social group to another (different uses, different behaviors, different perceptions and different relationships to technology): there are those who appropriate the technologies imposed on them (consumer products, uses imposed by law, etc.), those who make them their own by deconstructing them and diverting them from their primary purpose (hackers), and those who can adapt technologies to their needs (professional applications, in the army, for example). We can therefore, in a few formulas as lapidary as those in our tables, grasp what AI will or will not do, what it will or will not produce, because it is just as important to ask the question "who produces AI?", "which companies build AI?", etc.

Artificial intelligence will...	Date
"The achievement of above-human-level artificial intelligence will open to humanity an incredible variety of options."[40]	1976
"Research in artificial intelligence will increase our understanding of human language."[41]	1980
"Artificial intelligence will require new ways of thinking about computation that can exploit parallelism effectively."[42]	1981
"The widespread diffusion of artificial intelligence will have an impact on human society comparable to, if not greater than, the invention of the automobile or the printing press."[43]	1983
"Whether the aircraft is civil or military, artificial intelligence will have widespread application assisting the pilot in managing the systems involved."[44]	1985
"Artificial intelligence will change the workforce."[45]	2016
"Artificial intelligence will change the world."[46]	2017
"Artificial intelligence will make work more human."[47]	2018

40 http://jmc.stanford.edu/artificial-intelligence/reviews/weizenbaum.pdf.

41 https://pdfs.semanticscholar.org/cbe9/a097bde7cff080db81c5f4543f1847cf5384.pdf.

42 https://dspace.mit.edu/bitstream/handle/1721.1/6351/AIM-626.pdf?sequence=2.

43 https://fee.org/articles/japans-fifth-generation-computers-threat-to-the-free-market/.

44 https://apps.dtic.mil/dtic/tr/fulltext/u2/a164172.pdf.

45 https://www.canon.com.au/businessinsights/how-artificial-intelligence-will-change-the-workforce.

46 https://news.usc.edu/trojan-family/five-ways-ai-will-change-the-world-by-2050/.

47 https://timoelliott.com/blog/2018/11/why-artificial-intelligence-will-make-work-more-human.html.

Artificial intelligence will..	Date
"Artificial intelligence will benefit humanity."[48]	2018
"Artificial intelligence will impact business functions."[49]	2018
"Artificial intelligence will become weaponized in future cyberattacks."[50]	2018
"Like [...] electrical cables and computers before, artificial intelligence (AI) and big data will disrupt the way we work, live and even think."[51]	2018
"Artificial intelligence will change content marketing."[52]	2018
"Artificial intelligence will change the job market."[53]	2018
"Artificial intelligence will change every aspect of our lives."[54]	2018
"Artificial intelligence will change everything for patients and doctors."[55]	2018
"Will artificial intelligence revolutionize the art of war?" [NOË 18]	2018

Table 2.8. *Some assertions or assumptions about the future of AI*

Artificial intelligence won't/will not...	Date
"Artificial intelligence will not turn into a Frankenstein's monster."[56]	2014
"Artificial intelligence 'will not end the human race'."[57]	2015

48 https://www.inc.com/eric-mack/heres-27-expert-predictions-on-how-youll-live-with-artificial-intelligence-in-near-future.html.

49 https://www.inc.com/entrepreneurs-organization/4-ways-artificial-intelligence-will-impact-your-business-in-2019.html.

50 https://www.zdnet.com/article/this-is-how-artificial-intelligence-will-become-weaponized-in-future-cyberattacks/.

51 http://www.nas.gov.sg/archivesonline/data/pdfdoc/20180627005/Welcome%20Address%20by%20Defence%20Minister%20Dr%20Ng%20Eng%20Hen%20at%20Singapore%20Defence%20Technology%20Summit%202018.pdf.

52 https://www.forbes.com/sites/lilachbullock/2018/11/12/the-3-ways-that-artificial-intelligence-will-change-content-marketing/.

53 https://www.fool.com/investing/2018/10/25/3-ways-artificial-intelligence-will-change-the-job.aspx.

54 https://nypost.com/2018/09/07/how-artificial-intelligence-will-change-every-aspect-of-our-lives/.

55 https://www.zdnet.com/article/ai-in-the-nhs-how-artificial-intelligence-will-change-everything-for-patients-and-doctors/.

56 https://www.theguardian.com/technology/2014/aug/10/artificial-intelligence-will-not-become-a-frankensteins-monster-ian-winfield.

57 https://www.theguardian.com/technology/2015/jan/28/artificial-intelligence-will-not-end-human-race.

Artificial intelligence won't/will not…	**Date**
"True artificial intelligence will not exist until technology begins to think for itself."[58]	2017
"Artificial intelligence will not replace people."[59]	2018
"Artificial intelligence will not replace humans."[60]	2018
"Artificial intelligence will not kill jobs."[61]	2018
"Artificial intelligence will not replace the physician."[62]	2018
"Artificial intelligence will not replace human intelligence."[63]	2018
"Artificial intelligence will not be mainstream in marketing until 2021."[64]	2018
"Machines and artificial intelligence will not replace accountants."[65]	2018

Table 2.9. *Some assertions or hypotheses about which subjects will escape the impact of AI*

Artificial intelligence should…	**Date**
"The techniques of artificial intelligence should be applied to some realistic problems which exist in the programming and data processing fields."[66]	1971
"Advances in artificial intelligence should be incorporated into advanced computerized welding information services."[67]	1986

58 https://www.information-age.com/true-ai-doesnt-exist-augmented-intelligence-123468452/.

59 https://www.cnbc.com/2018/11/14/artificial-intelligence-will-not-replace-people-salesforceexecutive.html.

60 https://timesofindia.indiatimes.com/home/education/news/artificial-intelligence-will-not-replace-humans/articleshow/66215791.cms.

61 https://hackernoon.com/robots-and-artificial-intelligence-will-not-kill-jobs-instead-they-are-the-foundation-of-the-next-2e4c5c9348ff.

62 https://medium.com/datadriveninvestor/why-artificial-intelligence-will-not-replace-the-physician-e1bfd469743b.

63 http://www.theedgemarkets.com/article/artificial-intelligence-will-not-replace-human-intelligence-%E2%80%94-hsbc-report.

64 https://www.episerver.com/about/news/press-room/pressreleases/artificial-intelligence-will-not-be-mainstream-in-marketing-until-2021/.

65 https://www.rsm.global/singapore/news/why-machines-and-artificial-intelligence-will-not-replace-accountants.

66 https://dl.acm.org/citation.cfm?id=1622876.1622912.

67 https://www.gpo.gov/fdsys/pkg/GOVPUB-C13-6438b36d72e7975eac842f6dbcb25c82/pdf/GOVPUB-C13-6438b36d72e7975eac842f6dbcb25c82.pdf.

Artificial intelligence should…	Date
"Why artificial intelligence should terrify you."[68]	2017
"Artificial intelligence should be part of the recruiting process, but it can't replace the human touch."[69]	2018

Table 2.10. *Some assumptions about AI*

Artificial intelligence produces, or allows…	Date
"Artificial intelligence allows machines to reason and interact with the world."[70]	2016
"Data science produces insights; machine learning produces predictions; artificial intelligence produces actions."[71]	2017
"Artificial intelligence produces information."[72]	2018

Table 2.11. *Some consequences of artificial intelligence*

All these quotes highlight common views on artificial intelligence.

Its transformative power is central. Those who speak of it seem convinced that AI will have an impact on all human activities. It is still too early to measure the magnitude of these effects, even though, as is repeatedly stated, AI is not new and has existed for many years through various applications. The transformation is positive: it is defined as an advantage, an added value, a progress. This idea of "transformation" is hardly questioned except at the margin, when it is claimed that AI will not change… Potentially, therefore, transforming AI in all fields will have effects in the fields of security and defense. Some are affirmative (AI will revolutionize the art of war), others ask the question with more reserve (will AI transform the art of war?). These questions specific to the military field are

68 https://thebolditalic.com/why-artificial-intelligence-should-terrify-you-c9c891b94976.

69 https://ir.kornferry.com/news-releases/news-release-details/putting-ai-its-place-artificial-intelligence-should-be-part.

70 https://techcrunch.com/2016/10/23/advancements-in-artificial-intelligence-should-be-kept-in-the-public-eye/.

71 http://varianceexplained.org/r/ds-ml-ai/.

72 https://www.trudypeterson.com/blog/2018/3/13/commentary-artificial-intelligence-and-the-data-that-trains-it.

not unique to the last wave of AI. We saw in Chapter 1 of this book how much AI is linked to the military domain. So these questions have already been asked. It will be more interesting, however, to reflect on how the latest generation of AI (and its new promises of technological progress) renews – or not – thinking on security and defense policy, in strategic or operational fields. AI will be placed at the heart of new reflections on military transformation and revolution in military affairs. The cyber domain in particular will be impacted by this role of AI (cybersecurity and cyber defense).

The second major theme deals with the relationship between AI and humanity. For some, AI is an existential threat (which is reminiscent of the catastrophist, end-of-the-world discourse that accompanied the development of the Internet in the 1990s), for others it is not. Some see these technologies as new instruments of power, coercion, oppression and control, which will be used against humans; others oppose a more angelic vision, that of a society in which human and machine work together, where the machine is only an extension of the human being, to help people, accompany them and increase their capacities. A current of thought also tends to demystify and desacralize artificial intelligence, believing it would not be a miracle solution to problems, not as intelligent as is generally said and not as perfect, this train of thought highlights its lack of neutrality and its defects, in short its limitations.

2.4.2. *The absolute superiority of human intelligence over the machine*

One of the first objects of controversy lies in the recognition or denial of a possible equivalence, or even superiority, of the machine over the human. The debate lies in opposing or comparing the capacities of the human brain and those of the machine.

For those who are critical or skeptical about the prospects of AI, whose voices were heard at the dawn of the discipline in the 1960s, the sentence is defined: a computer will never be able to match a human in its activities. It may imitate a human, it may come close, but it will never do as well as human. It is not even a question of envisaging the machine surpassing a human. To support his criticism and his assertions, Hubert L. Dreyfus relies on observing advances in AI research over the period 1957–1967, which ended essentially in failure, none of the stated objectives having been achieved to date [DRE 72].

Many people, like H.L. Dreyfus, decided on a hierarchy that no progress can reverse: the human brain can never be surpassed in intelligence by the machine:

> "In *Le Cerveau et l'Ordinateur*, the aim is to demonstrate that, although competing, even surpassed in some areas by the electronic machine, the brain remains superior and the sole creator. [...] Does the native computer deserve the name artificial brain? Structural similarities cannot be denied. [...] But the computer will never reach a number of elements comparable to that of the brain. The brain as a whole has properties that the computer cannot achieve. [...] The performance of the computer [...] is unquestionably superior in the numerical domain, much less in the rational domain, not digital; it is inferior when it comes to intelligence, i.e. understanding, because intelligence is a global phenomenon that does not correspond to any particular nerve structure. From the box with the most learned 'expert system', only what is potentially contained can come out. The machine, unlike the brain, does not have a holistic production." [LAZ 88][73]

> "Machine intelligence will never match that of the human brain because it has no life. Artificial intelligence is the servant, the memory of human intelligence. There is no antagonism between them." [GIF 84]

2.4.3. *The replacement of humans by machines*

In *Computer Power and Human Reason* [WEI 75], Joseph Weizenbaum describes the replacement of the human being by the machine in tasks that should only be performed by humans as immoral. Science and technology provide us with tools, but their use is in some cases immoral. Computers, he said, are among those tools that impose moral considerations. The argument

73 Presentation of the work in the *Bulletin de l'Académie Nationale de Médecine*, vol. 172, no. 9, December 13, 1988.

has been met with diverse reactions [BUC 76], including being characterized as unreasonable [MCC 76].

Also expressed is the fear that humans will be overtaken by machines, by humanity's own creation, by our own creature: "Computer systems must not leave the field of algorithmic computing, so that humans can exercise control over them entirely." [WAR 80][74]

> "[These are] 'terrible prospects' according to a journalist from *Le Monde*. Don't machines that until now have only freed man from physical work, think they should replace him in the noblest tasks for which thought has so far ensured that man has a monopoly? Aren't these 'smart' machines, endowed with 'memory', now capable of surpassing the human brain in tasks that are being performed more and more every day? Isn't the supremacy of mankind threatened? Won't these 'thinking robots', in a future that some people predict, soon reduce humanity to slavery? […] For the former Minister of National Education, Edgar Faure, the 'machine that decides instead of executing' would be the root of a 'serious trauma for human intelligence'." [LAN 69].

But no machine, to counter these arguments and fears, will be able to totally escape the control, the will of humans, who alone decide on its implementation. In this sense, no machine can ever attain true autonomy:

> "An act is automatic if a machine performs it itself. […] But what machine acts without the participation of human desire? [...] Every machine remains inert outside the decision of man who puts it into action. […] It is therefore impossible to link, as has been done until now, the notion of automatism to the independence of the machine from man: the machine is always dependent on man." [DEL 53]

2.4.4. *AI as an existential threat*

The first level of threat lies in the creator's loss of control of the machine. This theme has been omnipresent since the early years of AI:

---

74 See the review by J.-M. Hoc, in *Le travail humain*, vol. 49, no. 3, pp. 280–283, 1986.

> "Some artificial intelligence specialists even believe that it is not impossible that machines can contribute to their own improvement, and that they will one day reach such perfection that we will become incapable of dominating them. This prospect is already causing unease among the younger generations, those who will be living in the year 2000. They fear mankind will be dehumanized." [BÉN 71]

In 2014, Stephen Hawking and Elon Musk took up the argument again, claiming that true AI could bring about the end of humanity. This position was nothing new and is part of the dichotomous reading of the impacts that technologies can have. The existential threat is one of the main themes of the *R.U.R.* play (1922) which introduces the term "robot":

> "The three initials stand for: Rozum Universal Robots. The constant development of mechanics and the growing evolution of modern machinery have enabled Karel Capek to look into the future, to make a foray into the world of tomorrow [...]. In this blessed time, it will be possible to produce man-machines, or, as the author says, Robots (a Czech word meaning workers)." [QUI 22]

The threat is built in the accumulation of events and facts. First of all, the robots "will be devoid of all human sensations, without the slightest notion of good and evil, since they are deprived of a soul […]." [QUI 22]

Then, as we unanimously agree that robots are useful, humans want to surround themselves with these machines. Karel Capek then imagines a world that closely resembles the Internet society: "We are in the offices of the director who dictates to his typist, Sylla, a voluminous letter. This chemically manufactured Robot is in charge of answering the numerous Robot orders from all over globe." [QUI 22]

Then, the robot enters the military domain [BLU 22]: "In the second act we learn that many governments have acquired Robots to wage war. Rozum has provided the world with this 'human' material."

Finally, the robot takes over humanity: "The systematic carnage completed, the Robots are engaged in a terrible massacre on all human beings who survived the war."

It was humankind who in reality laid the grounds for their own obsolescence.

For Ben Shneiderman, AI is one of the causes of the 10 plagues of the information age [SHN 86] (anxiety, alienation, unequal access to information, individual powerlessness in the face of organizations, overwhelming complexity and speed, organizational fragility, lack of professional responsibility, negative effects on employment, invasion of privacy and damage to the image of individuals):

> "With the presence of intelligent terminals, smart machines, and expert systems, it seems that machines have indeed taken over human abilities. These misleading phrases not only generate anxiety about computers, but also may undermine the image that we have of people and their abilities. Some behavioral psychologists suggest that we are little more than machines; some artificial intelligence workers believe that the automation of many human abilities is within reach. The rich diversity of human skills, the generative or creative nature of daily life, the emotional or passionate side of human endeavor, and the idiosyncratic imagination of each child seem lost or undervalued. Rather than be impressed by smart machines, accept the misguided pursuit of the Turing test, or focus on computational skills in people, the authors believe that we should recognize how designs that empower users will increase their appreciation of the richness and diversity of unique human abilities."

Several themes are intertwined here:

– super-intelligence, surpassing human intelligence, which would have the effect of surpassing humans and threatening them with death. This story is a fiction, a hypothesis excluded in the short and medium terms. It is not essential and does not ask immediate questions. However, it occupies part of the debates because it is sensational in nature. One can consider this theme as strong background noise;

– the effects of AI on societies and immediate realities: AI destroys jobs, but creates new jobs (what equilibrium?); AI at the service of States as an instrument of control and supervision;

– AI, the solution to our problems;

– the difficulties of science in achieving the goals set in the 1950s. Considerable progress has been made, but overall no AI is yet truly "smart".

AI needs to be demystified along with its current capabilities, what it really is compared to what it is or should be;

– the persistence of misunderstandings, misuse of concepts. Is ML really AI?

2.4.5. *The place of AI and robotics in fiction: the example of Japan*

In Japan, science fiction and manga have, for their part, given prominence to artificial intelligence in multiple forms (robotic or virtual) since the 1950s. AI and robotics often go hand in hand.

Literature makes it possible to distinguish:

– AI that play the main character;

– AI acting as an assistant, alongside the heroes or protagonists of different levels, who find themselves, thanks to AI, reinforced, helped and improved in often hostile or unknown environments; AI which is not the main protagonist; AI which is an object intervening in the story;

– purely virtual AI, which is located in cyberspace or in networks;

– AIs that are embedded in robots;

– embodied AI (it *is* a human body, is *in* a human body, or is biological).

We cross-reference these categories to put together Table 2.12, which is made up of narratives in specified manga (the list is of course not exhaustive. Japanese production has been impressive for several decades, and no catalogue or database lists them all).

	AI as the main protagonist	**AI as the hero's assistant, or as a supporting role, or as part of the narrative**
Virtual AI	*The Two Faces of Tomorrow* (1996, James P. Hogan, Yukinobu Hoshino): computers govern the world of the 21st Century. Artificial intelligence is built into the system, which acquires autonomous power.	*Sword Art Online* (2012–..., Reki Kawahara, Kôtarô Yamada): the Japanese government wants to create a high level artificial intelligence comparable to the human brain, to be integrated into weapon systems.

	AI as the main protagonist	AI as the hero's assistant, or as a supporting role, or as part of the narrative
AI incorporated in a machine, a robot	*Astro Boy* (1952–1968, Osamu Tezuka)[75]: the robot Astro fights crime and injustice. Astro is a superhero, who must save the world[76].	*Gundam Sentinel* (1987–1990, Masaya Takahashi): an AI named ALICE is integrated into the Gundam robot.
AI incarnate	*Doraemon* (1970–…, Fujiko F. Fujio): robot cat, friend of humans. *Pluto* (2003–2009, Naoki Urasawa)[77]: the Astro robot has an exceptional AI, capable of feeling human emotions. In this manga, we have robots/AI detectives, policemen, who fight against Pluto, the killer robot[78]. *Demokratia* (2013, 2015, Motorô Mase): the story of Maï, a female-looking humanoid robot whose actions are decided, via a social network, by 3,000 people.	*Biomega* (2004–2009, Tsutomu Nihei): the AI Fuyu Kanoe, integrated into the system of the hero's motorcycle, Zoichi Kanoe. AI is the hero's companion, helping him to analyze situations and giving advice on strategies. It is equipped with emotions.

75 This character influenced the youth of an entire generation, as did the robot cat, Doraemon. The Astro Boy robot is said to have played an important role in the development of Japanese intelligent robotics: "Many AI and robot researchers and engineers working successfully in Japan today say that they chose their career due to the wish to make their dream of creating Atom [Astro Boy] come true" [MAT 16]. "ASIMO's creator Masato Hirose said that he created the robot by imagining Astro Boy, one of the most famous Japanese anime characters and a robot. Hirose was ordered by Honda to create an Astro Boy, and he created ASIMO" [FUN 14].

76 These intelligent robots do good, fight against evil, are allies or friends of human. This benevolent look of men towards technologies that are at their service seems to have changed in the years 2010. There is now concern about the capabilities of AI and robots, their place in society and their negative effects [MAT 16].

77 Inspired by the manga *Astro, The strongest robot in the world,* by Osamu Tezuka (1964).

78 For a presentation of the seven robots that are present in the story, the reader can visit: https://labasesecrete.fr/comparatif-entre-pluto-et-astro-le-robot-le-plus-fort-du-monde/.

	AI as the main protagonist	**AI as the hero's assistant, or as a supporting role, or as part of the narrative**
AI incarnate	*Atom: The Beginning* (2014–…, Tetsuroh Kasahara): story of the creation of the robot A106 (or "six") with individual consciousness. *High School Prodigies Have It Easy Even In Another World* (2015– …, Riku Misora, Kôtarô Yamada): Chapter 35 entitled *"Artificial Intelligence"* has Ôhoshi Ringo as its protagonist. She is a creature of flesh and bone artificially created by genetic manipulation. She is far more intelligent than all humans: "it has been said that five years of my work had advanced humanity by five hundred", says the heroine.	*Ghost in the Shell* (1989, Masamune Shirow): the manga will have two sequels: *Ghost in the Shell 2: Man–Machine Interface* and *Ghost in the Shell 1.5: Human Error Processor*. The main character, Motoko Kusanagi, is a female cyborg, a police officer. Her role is to hunt down a cybercriminal, who is actually an AI who has developed consciousness and has incarnated into an android.

Table 2.12. *Some AI and robots in Japanese manga and anime*

2.5. Political considerations

Harry Killer built a town called Blackland. We are here in a fictional story, imagined by Michel Verne (son of Jules Verne), in his novel *L'étonnante aventure de la mission Barsac*, published in 1919 [VER 19][79]. This city is structured in three parts separated from each other by high impenetrable walls. The inhabitants of the aristocratic quarter are organized in a military manner. Their function is to wage war. They also have a police function. A third district is inhabited by white people who have not yet been able to access the first one. Their function is that of commerce. Between the two districts is the slave district. The rulers have taken up residence in a town

79 In spite of it being written by Michel Verne, the novel was published under the name of Jules Verne.

a little far from Blackland and have their palaces there. Blackland is equipped with the latest technological innovations, and no quarter, aristocratic or slave, is without them: "No home [...] that didn't have a telephone. Not a street, not a house, not even a hut in the slave quarter that did not have pressurized water and electrical lighting."

On one of the towers overlooking the city, Harry Killer, the despot leader, had a machine built to monitor the actions of the population:

> "Thanks to its optical disposal, this instrument, which has been given the name 'cycloscope', seems to straighten vertically this circular strip of land, whose care-taker, who stands at the center of the device, constantly has all the points under his eyes magnified [...] the outer world changes its appearance in our eyes. From whichever side they are facing, we first see only a vertical wall, a network of black lines divided into a multitude of distinct little squares [...]. This wall, whose base is separated from us by an abyss of darkness, and whose summit seems to rise to a prodigious height, seems to be made of a sort of milky light. However, we soon find that its color is far from uniform, but that it is, on the contrary, the result of an infinity of spots of different tones, with rather indecisive contours. A moment's attention shows us that these spots are some, trees, others, fields or paths, others still, men working the earth, the whole being sufficiently enlarged to be recognized without effort [...]. 'You see these negroes', says Harry Killer, pointing to two of the spots in question, which are separated by a large interval. 'Let's suppose they have the idea of running away. It wouldn't take long.' While talking, he grabbed a telephone transmitter. 'One hundred and eleventh circle; radius fifteen hundred and twenty-eight', he said. Then, seizing another transmitter, he added 'Fourteenth circle. Radius six thousand four hundred and two.' Finally, turning towards us, 'Take a good look', he recommends. After a few moments of waiting, during which nothing special happens, one of the spots is suddenly masked by a cloud of smoke. When the smoke has cleared, the stain has disappeared. 'What happened to the man who was working there?' 'He's dead', Harry Killer answered coldly. 'This one was demolished by an aerial torpedo. It's a kind of rocket that carries up to twenty-five kilometers, and whose speed and precision you could appreciate.'"

In this fictional story, Michel Verne describes:

– a “modern” city, technologized, equipped with modern means of communication;

– panoptic surveillance based on a combination of:

- despotic power,

- ability to draw some of its power from the mastery of high-tech technologies, in this case surveillance technologies, consisting of vision instruments, imaging, a screen that reflects an image of the world, making surveillance more effective because every object, every individual, is located there, has its coordinates in space; this visualization instrument used for surveillance is coupled with a communication system, a C2 military center;

– human targets that are individually neutralized with missiles. The motive for their elimination can be cruel – the despot wants to show his power of life and death – and punishes any deviant behavior (attempted escape, breaking into forbidden spaces, etc.).

As is very often the case with anticipatory novels, we can see analogies with our contemporary world:

– the modern city (smart city, hyper-connected city) which makes its hectares benefit from technological innovations;

– monitoring systems are deployed throughout the world;

– satellites, drones, cameras (surveillance, on board, police, military, etc.) capturing the world in images; screens are omnipresent and are the essential link in any modern surveillance system;

– military command systems are designed around machines processing images; objects in the images are identifiable but often, due to the degradation of quality in night or difficult environments (– for example, combat situations, images taken from too great a distance, etc.) – the objects in the images are not identifiable. Instead they are often only imprecise forms that are displayed and on the basis of which the command or the soldier must interpret, analyze and decide. The sentence “one of the spots is suddenly obscured by a cloud of smoke” refers us to images of videos taken from combat helicopters or drones when targets are engaged, neutralized; targets – human or material – are nowadays individually marked out, according to the principle of precise surgical strikes.

This system has its modern equivalent, in the age of computers, satellite imagery, drone images, omnipresent surveillance cameras, mobile telephony that makes it possible to trace the movements of individuals, an Internet that makes it possible to track their habits, opinions, facts and gestures. Modern systems, to distinguish between tasks, to differentiate between objects and human beings, include artificial intelligence, pattern recognition applications, as well as faces. In war, we can imagine systems that combine these recognition tools with lethal capabilities. We reproduce the principles that appear in the system described by Jules Verne: power, the power of life and death, based on surveillance through the exploitation of images, and the physical elimination of disruptive elements (as we can see in wars or the fight against terrorism, but as we can also imagine in a totalitarian and police state).

One of the themes most frequently associated with artificial intelligence is its impact on societies and on the world as a whole. The Brookings Institution has entitled one of its publications "How artificial intelligence is transforming the world." [WES 18]

These major effects are expected to occur at different levels of society: on the economy level, on the employment level [SMI 14] and many others.

But if AI makes possible a scenario such as the one described by Michel Verne, it is not only an instrument of power, a "strong arm".

The security and defense sectors are also benefiting or will be affected by AI-induced changes.

Faced with the diversity of known, envisaged and yet to be invented uses of AI, faced with the potentialities of AI, real or imagined, governments have taken up the subject, producing "national strategies" for AI, for research in AI, for security, for the industrialization of AI, displaying all their ambitions in these documents. These strategies or programs are focused on economic, social, research-related and industrial issues. Security and defense issues are also reconsidered in the light of artificial intelligence.

2.5.1. *National strategies for artificial intelligence*

A national strategy can meet several expectations: coordinate actors and actions, respond to the expression of a need for coordination, respond to a broad societal issue, define a precise framework for the action of an ecosystem of actors on specific issues, ensure a better use of resources and optimize resources that are too scattered. Deciding on a national strategy therefore means drawing the broad outlines of a policy for a given period, managing resources, wanting to optimize by coordinating resources at national level in order to obtain better results (there are regional or international strategies, the European Union for example). A strategy may decide to create new organizations or institutions, it opens the way for action, allows actors in a field to organize themselves, to mobilize, to act with reference to the strategy defined by the State. It defines the framework for action.

A strategy must therefore *a priori* define its objectives. The findings that motivate it must make it necessary and urgent: "In the past, certain criticism could be heard […] regarding the current quality of investigations and proceedings for war crimes […] The lack of […] An unsatisfactory level of […] Poor quality of […] Due to the above, it is necessary to adopt a strategy to prosecute war crimes at the national level." [REP 16]

Many expectations are sometimes placed in a national strategy: "a strategy […] whose objectives and activities would provide a comprehensive and clear strategic framework for improving all areas where problems have been identified." [REP 16]

Title of the document	Country	Author	Year
"Artificial intelligence: opportunities and implications for the future of decision making"	United Kingdom	Government Office for Science	2015
"Three-Year Guidance for Internet Plus Artificial Intelligence Plan (2016–2018)"	China		2016
"The national artificial intelligence research and development strategic plan"	United States	National Science and Technology Council. Networking and Information Technology Research and Development Subcommittee (NITRD)	2016
"Artificial intelligence for Africa"	South Africa	Access Partnership. University of Pretoria	2017

Title of the document	Country	Author	Year
"Pan-Canadian artificial intelligence strategy[80]"	Canada	CIFAR, funded by the Canadian government, Facebook and the RBC Foundation	2017
"A New Generation of Artificial Intelligence Development Plan"	China	Chinese State Council	2017
"Artificial Intelligence Program"	Finland		2017
"Artificial Intelligence Technology Strategy"	Japan	Report of Strategic Council for AI Technology	2017
"AI Singapore"	Singapore	Government	2017
"Artificial Intelligence strategy"	Germany	The Federal Government	2018
"Three-Year Action Plan for Promoting Development of a New Generation Artificial Intelligence Industry (2018–2020)"	China		2018
"2018 White House Summit on Artificial Intelligence for American Industry"	United States	White House	2018
"Why the United States Needs a National Artificial Intelligence Strategy and What It Should Look Like"	United States	Center for Data Innovation	2018
"Cédric Villani Report"	France	Congressman. Villani Mission	2018
"#FranceAI Strategy"	France	Agoranov, Secretary of State for Innovation, Secretary of State for Higher Education and Research	2018
"National Strategy for Artificial Intelligence #AIFORALL"	India	NITI Aayog	2018
"AI Masters Program"	Ireland	Private Sector Initiative	2018
"AI Task Force"	Italy	Government	2018

80 AI Research and Innovation Program. This strategy is not the official policy of the Canadian government, but a project of CIFAR, a research association, whose strategy formulated here responds to a government request. This strategy is defined as a project for academic research and development in the field of AI.

Title of the document	Country	Author	Year
"Blockchain & Artificial Intelligence task force"	Kenya	Government	2018
"Lithuanian artificial intelligence strategy. A vision of the future"	Lithuania	Ministry of Economy and Innovation	2018
"Mauritius Artificial Intelligence Strategy"	Mauritius	Working Group on AI – Chairmanship of the Secretary to the Cabinet	2018
"Strategy for the Development of Quebec's Artificial Intelligence Ecosystem"	Quebec	Artificial Intelligence Cluster Steering Committee. University of Montreal (government-mandated)	2018
"National Approach to artificial intelligence"	Sweden	Ministry of Enterprise and Innovation	2018
"Beijing AI Principles"	China		2019
"National Strategy for artificial intelligence"	Denmark	Ministry of Finance and Ministry of Industry, Business and Financial Affairs	2019
"Building a World-Leading AI and Data Strategy for an Inclusive Scotland"	Scotland	Scottish Council for Development and Industry	2019
"Spanish RDI Strategy in Artificial Intelligence"	Spain	General Technical Secretariat of the Ministry of Science, Innovation and Universities	2019
"Kratt report"	Estonia	Government	2019
"Intel's recommendations for the U.S. National Strategy on Artificial Intelligence"	United States	Intel	2019
"Leading the way into the age of artificial intelligence: Final report of Finland's Artificial Intelligence Program 2019"	Finland	Publications of the Ministry of Economic Affairs and Employment	2019
"AI for Humanity"	France	Government	2019
"Towards an AI Strategy"	Malta	Office of the Prime Minister	2019

Title of the document	Country	Author	Year
"AI Portugal 2030"	Portugal	Ministry of Science, Technology and Higher Education	2019
"National artificial intelligence strategy for Qatar"	Qatar	Qatar Center for Artificial Intelligence (QCAI). Hamad Bin Khalifa University	2019
"National Artificial Intelligence Strategy of the Czech Republic"	Czech Republic	Ministry of Industry and Trade	2019
"Decree of the President of the Russian Federation on the Development of Artificial Intelligence in the Russian Federation"	Russia	Government	2019

Table 2.13. *National AI strategy documents*

Artificial intelligence has emerged as a topic requiring the development of national strategies. A review of these strategies shows that there has been an increase in the number of national strategies since 2015. However, not all these documents considered as national strategies are produced by governments. Their authors include research centers and private actors. Among State authors, the strategies are written by various ministries: the economy, security, defense, etc. However, the predominance of the ministries in charge of the economy and industry should be underlined. Artificial intelligence seems to be, first and foremost, an economic issue for the States. Contributions for society or humanity in terms of knowledge production, dissemination of knowledge, improvement of life conditions therefore lie far from these considerations. The primary objective is the management of an economy, the development of an industry, certainly based on R&D, but above all by supporting trade.

All of the national strategy documents we refer to here come from a variety of governmental and non-governmental sources (as is the case with documents for India, Quebec and Africa, for example).

From these "strategies", we will retain the main stories and their constituent elements. What are these strategies about? What visions of the world do they propose and how do they propose to build societies that consider artificial intelligence?

For the majority of States, AI is a high value-added, creative, transformative technology. And if the State does not ignore the few "negative" aspects of these technologies (its risks, its dangers), it seems that it would be sufficient to counter them by a few frameworks, by law in particular, by ethical control, to control their effects and leave room for the positive dimension (among the tools for reducing the negative dimension are: law, standards, education, training, awareness, control of practices and users. Resilience is also discussed).

Supported by scientists who sell or oversell the benefits of AI, albeit while pointing out their negative characteristics, politicians construct a discourse depicting an AI that is not only beneficial, but necessary and unavoidable. Progress, the economy, social change and social justice (AI at the service of the fight against fraud, for example) are evoked, so too are higher, even indisputable interests, such as security (AI that predicts, anticipates or detects terrorists, criminal projects) or power on the international scene.

	Argument, theme	**Examples**
Emergency	AI is progressing rapidly and bringing about profound and rapid changes: we urgently need to take this new reality into account.	German Strategy 2018
Transformative power	AI has a strong potential, it is an innovation with deep *transformative* power. But it is still a *mature technology, albeit still rudimentary.*	Strategy for Africa, 2017: "AI will be the largest technology revolution of our times and has the potential to disrupt almost all aspects of human existence" Quebec Strategy, 2018: "Artificial intelligence (AI) is poised to transform Quebec's economy in a radical way"
Key factors, determinants	The country must invest in[81] AI so as to not miss the technical, economic and social shift.	German Strategy 2018

81 The history of AI is marked by its industrialization and commercialization, which began in the 1970s. The discipline therefore quickly moved out of universities and research centers, to win over industry and become a new market, and found itself at the heart of the technological war that pitted Japan against the West in the 1980s. In 1982, Japan launched its "Fifth Generation Computer Systems" project, which aimed at technological domination in the field of computer science, of which AI formed part. In response, the United States included AI in DARPA's new 10-year development program. AI is also included in the projects of the

	Argument, theme	Examples
Key factors, determinants	The promotion of AI is a collective affair, part of an *ecosystem.*	Strategy for Africa, 2017; Quebec's Strategy, 2018
	AI must be thought of as a whole, not only in terms of research, applications and economics, but also in terms of its social implications (legal, ethical and political).	2017 Pan-Canadian Strategy
	Research and innovation are essential.	2017 Pan-Canadian Strategy; German 2018 Strategy
	The country has been investing in AI for several decades. The relationship with AI is therefore anchored and cultural, the present has a real anchoring. The *historical presence of high-level R&D* is the starting point for any development of an industrial, economic and social *ecosystem.*	2017 Pan-Canadian Strategy: "For more than 45 years, Canadian researchers have been at the forefront of advancing artificial intelligence (AI) research" German Strategy 2018: "Germany is very well positioned due to its excellent, broad-based research landscape" Quebec Strategy, 2018: "Quebec has experienced a historic boom in artificial intelligence, thanks to the excellence of academic research in the field that has led to the arrival on the scene of technology giants"
	The R&D community needs to be energized and animated through symposiums, training programs and so on.	2017 Pan-Canadian Strategy
"Positive" AI	The development of AI in society cannot take place without *respect for fundamental values* (freedoms, fundamental rights), law, ethics, culture, etc. We must be vigilant and create the conditions for this protection. AI must work for the good of humanity and the environment.	German Strategy of 2018; Beijing AI Principles, 2019

Microelectronics and Computer Technology Corporation (MCC), a consortium of some 20 companies led by former NSA director Bobby Inman. Although AI is not the only object of

	Argument, theme	Examples
"Positive" AI	AI as an *opportunity for…* economy, society, growth, democratization…	Strategy for Africa, 2017
	AI as a "solution" to challenges, problems… in various fields (health, agriculture, finance, public services, reducing poverty, unemployment, improving education, fighting epidemics, improving food productivity, ensuring growth in key sectors, etc.).	Strategy for Africa, 2017: "Artificial intelligence (AI) technologies have the potential to solve some of the most pressing challenges that impact Sub-Saharan Africa"
	Use of AI should be for the *benefit of* all citizens (employment, climate, economy, etc.).	2018 German strategy: "AI which serves the good of society"
The State, its role, its responsibilities	The *State* creates a framework and regulates.	Strategy for Africa, 2017
	The *State* must act as an advocate and protector of rights and values, as citizens are not adequately trained and informed about the impacts of AI on society.	German Strategy 2018
	The *State* invests.	German Strategy 2018 (€500 million in 2019: €3 billion by 2025)
	The *government investment* will not be made at a loss.	German Strategy 2018: "The leverage effect this will have on business, science and the Länder (Federal States in Germany) will mean that the overall amount available is at least doubled"

these international clashes around technologies, it is one element. Read: *Cryptolog* magazine, March–April 1986, NSA, USA, 16 pages, https://www.nsa. gov/Portals/70/documents/news-features/declassified-documents/cryptologs/cryptolog_101.pdf.

	Argument, theme	Examples
The State, its role, its responsibilities	Involve the *State*, companies and society in exchanges, dialogues and discussions for harmonious development that respects the rights and interests of all.	German Strategy 2018
	The national *government* must support the dissemination of AI in the economic fabric, in industrial activity.	German Strategy 2018
Risks, threats: preventing, anticipating, empowering	Caution, reserve. AI is not without *risks*: increased concentration of power in the hands of a few (predictive abilities, control over individuals via control over data, etc.), massive destruction of jobs (which must therefore be transformed), leaving entire sections of the population (unconnected, untrained citizens, etc.) out of the loop, blind confidence in systems whose data may nevertheless be of poor quality and whose algorithms may be tainted by bias (coding discrimination in automated decision-making systems), etc.	Strategy for Africa, 2017: "Machine learning processes are only as good as the data that are fed to them. This is why biases, deeply-rooted in human society, can and do creep in. AI algorithms integrate biases inherent in data, or even in the person who built the process, further perpetuating existing societal biases" Beijing AI Principles, 2019
	Responsibility of researchers to think about the ethical, social, etc. implications of their inventions.	Strategy for Africa, 2017; Beijing AI Principles, 2019
	Preventing the *risks* of AI upstream: for example, developing algorithms by mobilizing engineers/researchers of different origins (cultures, politics, nationality, sex, etc.) to *avoid bias*.	Strategy for Africa, 2017

	Argument, theme	Examples
International	AI is a key element in the positioning of the *national economy in the global system.*	German Strategy 2018
	The country or regional area must become a leader, a major center in the world.	German Strategy 2018
	Internationalization: AI cannot be thought of, promoted and developed within a single national framework.	In Europe, national strategies are defined from a European perspective. Action must be collective: German Strategy for 2018 A Strategy for Africa, 2017, considers AI at the scale of sub-Saharan Africa
	R&D capacity must be concentrated around large, world-class institutes of excellence.	2017 Pan-Canadian Strategy
Data	AI (the ML) is dependent on the quantity and quality of the data. *Access to the data must be ensured*, the data must be massive, comprehensive, representative (not risking ignoring entire sections of the population).	Strategy for Africa, 2017

Table 2.14. *Key elements of the narratives of national strategies for artificial intelligence*

Several accounts consider AI to be an unknown: its technological development is rapid, gaining speed from States and societies that are not yet prepared for it. No one is sure where all this may lead societies that do not yet have a true AI culture. If the general movement that societies must make seems inevitable, if only for economic reasons, it must be undertaken without delay but with the necessary precautions to ensure that the effects are not perverse. As was the case with the strategies for a digital society or economy or the information highways projects in the 1980–1990s and following years, these strategies make the new AI technologies the vectors of almost universal economic growth and social progress, with the exception, however, that there are many reservations or reservations about AI. Governments want to reassure societies, showing their willingness to frame developments and uses of technology to protect against the potential risks of

AI, which to date have not been very real. We are of course entitled to consider that this discourse is in denial or a lie, because we can very well consider the potential risks of AI for the fundamental rights of citizens: when AI is intensively exploited for security and defense purposes, when citizens are traceable and controllable, when their actions and gestures can be analyzed and reconstructed, when nothing in their lives can escape the scrutiny of the police state, then the rights of citizens are indeed eroded. Communication interception programs, mass surveillance programs in the United States and many other States, at different scales, have already, in the past few years, shown the range of possibilities and, above all, the intentions of State actors in their social control practices. These practices have already been able to take advantage of AI's capabilities and will do so even more in the future. But this dimension does not appear in strategic narratives, as they are formulated by governments.

Strategies in many ways are as much like specifications as they are like professions of faith: "We will develop... we will ensure... we will monitor... we will work together..."

Many speeches are also idealistic. For if all States claim either to have the best potential in the world or to develop the best potential, to be a global hub or "the" global hub, it is obvious that not all will succeed. Leadership positions are rare. Specializations will therefore be needed to claim eventual leadership in niches within the broad field or sector that is AI.

The promotion of initiatives, State support for industry, research and development, i.e. the massive investment required by the very wide range of declared initiatives and promises, will come up against economic and budgetary realities. These commitments will be delivered at the expense of other policies, and will only be delivered if economies are doing well. The development of AI is dependent upon the fate of State budgets.

Also noteworthy is the relative confidentiality in which the relationship between AI and national security, security issues generally, is maintained in these documents. The latter are certainly more general in scope and, like their predecessors, which are plans, programs, initiatives or strategies for the development of information highways, the Internet mainly addresses the general conditions for the appropriation of these technologies by societies. Security is undoubtedly a particular issue. However, security may appear in some documents that refer to the many areas that will benefit from AI (in the fight against financial delinquency, for example, which is an aspect of security policies). But overall, the fight against crime or cybersecurity is not highlighted in all these documents.

2.5.1.1. *Russia's strategy for 2019*

On October 10, 2019, the Russian government published its national strategy for AI. The document is entitled "Decree of the President of the Russian Federation on the Development of Artificial Intelligence in the Russian Federation"[82]. The strategy formulates medium-term (2024) and longer-term (2030) objectives. This decree also results in an amendment to the "Digital Economy of the Russian Federation" program before the end of 2019. The document proposes the official definition of AI:

> "Artificial intelligence – a set of technological solutions that makes it possible to simulate human cognitive functions (including self-learning and seeking solutions without a predetermined algorithm), as well as to obtain results during the performance of specific tasks that are at least comparable to the results of human intellectual activity. This set of technological solutions shall consist of information and communications infrastructure, software (including that in which machine learning techniques are employed), and data-handling procedures and services."

This is a very pragmatic definition: AI must provide solutions, results.

The national project officially places the development of AI under certain constraints: development respectful of freedoms, human rights, the right to work, the obligation to minimize risks in the face of negative consequences of AI, transparency (of data processing, algorithms, AI processes), national sovereignty (Russia must develop its own technologies). AI is assigned several goals, which must serve the economy, politics, national security, government and other purposes.

To ensure the development of AI, Russian policy will focus on supporting scientific research, software development, quality data production, training (for jobs requiring the use of AI), stimulating investment, attracting investors, and building favorable economic conditions.

---

82 English translation available at: https://cset.georgetown.edu/wp-content/uploads/t0060_Russia_AI_strategy_EN-1.pdf.

The terms "cyber", "military", "defense" and "army" are absent from this document, which therefore does not deal with the national security and defense dimensions.

2.5.2. *U.S. policy*

2.5.2.1. *Barack Obama's presidency*

During President Barack Obama's term, the U.S. authorities published several documents and launched several initiatives for the development of AI.

In May 2016, the "White House Future of Artificial Intelligence Initiative" was implemented, which would involve five workshops and a short public consultation (Request for Information) (June 27 to July 22, 2016, less than a month).

All 161 responses to the consultation were compiled in a 349-page document [WHI 16], published in September 2016. It should be noted that the responses were not anonymous. Although the number of responses received may seem small and may not reflect the opinion or mindset of the overall U.S. population regarding AI, the content of the responses is informative. The word cloud (based on the 65 most commonly used terms in all the texts) highlights a few major themes: research, people, systems, machines and technologies, the public, and the government. Reflections thus revolve around the relationship between humans and machines, and the role of research and government. The concepts of security, insecurity, risk, threat, war, military, attack and industry are absent from the primary concerns.

– The individual, the human being, society: society, the contributors recall, must be informed of the uses made of the citizens' data. For if these uses can be legitimate and in the interest of society (AI as a "public good") in order to provide new services, improve living conditions and benefit everyone, it is also important to ensure that the State does not cross a red line ("we need to see clearly how easy it can be to cross over from public good to control with unintended consequences and impact on citizens' human rights"; "the use of AI for public good: how to use AI to catch the 'bad guy' without turning our country into a police state?").

– The notion of a "public good" (which is one of the themes proposed in the survey terms of reference) is subject to interpretation. Some contributors are very cautious ("'The Public Good' is open to interpretation") and others are very critical of this concept which the authorities have used in a questionable manner ("There are many problems with using either AI or government action to achieve a goal of 'public good' when part of the program includes actively harming some people", "In the past, things that were deemed part of the 'public good' included slavery, poll taxes, keeping black people away from white water fountains, not letting women vote, etc.").

– The notion of "control" is common in the debates. Contributors insist on the necessity and the difficulty of "controlling" AI systems, but control is a strictly technical notion (ensuring that the system is secure, that it is not hacked, that its data is not altered, ensuring that the control of systems or machines called "automatons" is maintained) and does not create the challenges of police control or social control, for example.

Figure 2.12. *Cloud of the 65 most commonly used terms in the OSTP Request for Information on the Future of Artificial Intelligence (2016) [WHI 16]. For a color version of this figure, see www.iste.co.uk/ventre/artificial.zip*

Two official publications were to be produced in October 2016 to give concrete expression to all the reflections carried out in the framework of the "White House Future of Artificial Intelligence Initiative", or as part of their prolongation.

Title	Date	Source
"White House Future of Artificial Intelligence Initiative"	May 2016	White House
"Request for Information on the Future of Artificial Intelligence. Public Responses"	September 2016	White House
"Preparing for the future of artificial intelligence"[83]	October 2016	Executive Office of the President. National Science and Technology Council Committee on Technology
"The National artificial intelligence research and development strategic plan"[84]	October 2016	National Science and Technology Council Networking and Information Technology Research and Development Subcommittee

Table 2.15. *U.S. AI initiatives and strategy papers released in 2016*

In "Preparing for the future of artificial intelligence" (2016), the dominant terms are data, technology, future, development, science, research and security (safety). This nearly 50-page document is built around the following topics:

– definitions and background of AI;

– AI applications for the good of society;

– the regulation of AI;

– research and human resources;

– the economy;

– safety and governance;

– security, particularly cybersecurity and the militarization of AI (AI in weapon systems).

One of the priorities in cybersecurity is the development of AI tools to autonomously detect, assess and patch software vulnerabilities before hackers exploit them. The report mentions the DARPA's Cyber Grand Challenge

83 https://obamawhitehouse.archives.gov/sites/default/files/whitehouse_files/microsites/ostp/NSTC/preparing_for_the_future_of_ai.pdf.

84 https://obamawhitehouse.archives.gov/sites/default/files/whitehouse_files/microsites/ostp/NSTC/national_ai_rd_strategic_plan.pdf.

(CGC) which is pursuing this objective and would have achieved interesting results.

On defense, the report recalls the American position: the United States will develop autonomous smart weapons only in accordance with the international humanitarian law, including the principles of distinction and proportionality.

Figure 2.13. *Cloud of the 65 most commonly used terms in the report "Preparing for the future of AI" (2016). For a color version of this figure, see www.iste.co.uk/ventre/artificial.zip*

The document "The national artificial intelligence research and development strategic plan" (2016) focuses particularly on R&D strategies. As will be the case for the majority of official texts published in the world, AI is presented as a breakthrough technology, heralding profound transformations in all areas of human activity, for the good of society and economic development: "Artificial intelligence (AI) is a transformative technology that holds promise for tremendous societal and economic benefit. AI has the potential to revolutionize how we live, work, learn, discover, and communicate."

The objective is to develop a positive, beneficial AI and to minimize its negative aspects as much as possible. Implementation of the proposed strategy is based on seven priorities: ensuring long-term investment; promoting AI that supports people and does not merely replace them; taking into account the ethical and legal aspects; securing AI systems; making quality databases available to researchers and AI developers and facilitating their access to them; carrying out technology assessments to define standards; assessing the needs in terms of human resources to develop AI.

AI for economic prosperity remains a key component of this policy.

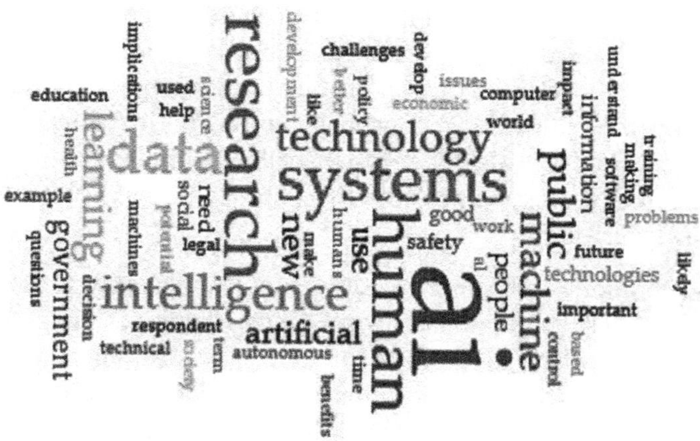

Figure 2.14. *Cloud of the 65 most commonly used terms in "The national artificial intelligence research and development strategic plan" (2016)*[85]*. For a color version of this figure, see www.iste.co.uk/ventre/artificial.zip*

2.5.2.2. *The presidency of Donald Trump*

Continuing the approach of previous governments, Donald Trump's administration is conducting new consultations and publishing new strategy documents. The National Institute of Standards and Technology (NIST) launched a consultation in 2019, and all responses, which are not anonymous, are also available online[86]. The term is plural. Major companies are taking part in the debates (IBM, Microsoft, Google, Amazon, etc.). As the responses are not constrained by a particular format, their developments vary from a few lines to a few dozen pages, with contributors sending either

85 Illustration created using the online application: https://voyant-tools.org/.

86 Comments Received for RFI about Federal Engagement in Artificial Intelligence Standards, 2019, available at: https://www.nist.gov/topics/artificial-intelligence/comments-received-rfi-about-federal-engagement-artificial.

simple comments, reports or even scientific articles. From this diversity, we will, however, retain some of the issues that emerge:

– Building confidence in AI-using technologies: one comment[87] criticizes the common discourse that: "a trustworthy system is a technically correct system". Trust is not reducible to a technological or normative issue. There is an essential social and political dimension: respect for freedoms, privacy and the informed consent of individuals must be guaranteed. AI can also often resemble a black box that is incapable of validating the results it produces. "Today's crisis of trust in artificial intelligence (AI) stems from a perception on the part of operators that they cannot explain why a given result was generated […]. In order for algorithms to be trusted, they must be explainable – i.e. it must be possible for a human to make sense of why an algorithm did what it did."[88]

– The uncertainties, which are great around the technology, the quality or reliability of its results, its impact in the fields where it is integrated: "there is no national standard to evaluate or otherwise examine the success or failures of artificial intelligence within a given context, process, or sector […] in medicine, medical professionals have zero risk tolerance because of what is at stake – their patients' health, their reputation, their license to practice. Medical professionals would not adopt A.I. without proof of credibility, accuracy, precision, or similar reputable, standardized metrics."[89]

– The notion of "risk" is ubiquitous, referring to technological risks, risks to citizens' rights, values, ethics, business, economy, democracy.

– Considerations for privacy, freedoms, fundamental rights, citizens' rights, ethics and the risks weighing on all these variables are a strong constant in all responses and more globally for the human being (the risks weighing on individuals, taking into account disabilities, "negative impacts on socioeconomically disadvantaged groups" [STU 19], etc.).

---

87 Maintaining American Values with AI Ethics Standards for Data Collection and Algorithmic Processing, available at: https://www.nist.gov/system/files/documents/2019/06/03/nist-ai-rfi-lchin_001.pdf.

88 Comments in response to the National Institute of Standards and Technology Request for Information on Developing a Federal AI Standards Engagement Plan, June 7, 2019, available at: https://www.nist.gov/sites/default/files/documents/2019/06/07/nist-ai-rfi-broniatowski-caliskan-reyna-schwartz.pdf.

89 Buoy Health, Inc., RFI: Developing a Federal A.I. Standards Engagement Plan, June 7, 2019, available at: https://www.nist.gov/sites/default/files/documents/2019/06/10/nist-ai-rfi-buoy-health-inc-001.pdf.

– The creation of norms or standards for AI is an international issue, in which the major powers defend their national interests and want to dictate their rules by claiming a leadership status: "The U.S. has an opportunity to lead the way in defining guidelines, regulations, and laws that address ethical issues in AI."[90]

– State/private balance: the development of AI raises the question of the State/private sector ratio, of the balance to be ensured between the two categories. Should the State really intervene to define the standards of AI? Isn't the private sector better equipped for defining this? "The U.S. private sector currently leads in AI research and development, but requires federal support to engage successfully at international standards bodies."[91]

– China is repeatedly referred to as "negative": "China has a history of subverting international technical standards, such as by developing alternative national standards, to create trade barriers that favor its domestic interests" [CDI 19]; "Chinese oppression of the Uighur minority population in Xinjiang has been built largely through the rise of AI-enabled technologies" [CHI 19]; "The Chinese Social Credit System is at odds with U.S. perspectives on the values of liberty and freedom as well as other countries' values on personal data privacy and protection as seen with the GDPR."[92] Here, we see a reappearance of the fundamental disagreements that have been expressed in international debates on Internet governance around AI issues. There is a dividing line between the United States and China, as well as Russia: "It would be problematic to start an AI race with China or Russia, as those nations have very different views of privacy, security and liberty than the U.S."[93]

90 RTI International, Response to NIST RFI 2019-08818: Developing a Federal AI Standards Engagement Plan, 2019, available at: https://www.nist.gov/sites/default/files/documents/2019/06/10/nist-ai-rfi-rti-001.pdf.

91 Center for the Governance of AI, the Future of Life Institute, the Center for Long-Term Cybersecurity, RFI: Developing a Federal AI Standards Engagement Plan, June 6, 2019, available at: https://www.nist.gov/sites/default/files/documents/2019/06/06/nist-ai-rfi-fhi-fli-cltc-cfi-001.pdf.

92 RTI International, RFI: Developing a Federal AI Standards Engagement Plan, June 2019, available at: https://www.nist.gov/sites/default/files/documents/2019/06/10/nist-ai-rfi-rti-001.pdf.

93 Hitachi Vantara Federal (HVF), RFI: Developing a Federal AI Standards Engagement Plan, May 31, 2019, available at: https://www.nist.gov/sites/default/files/documents/2019/06/11/nist-ai-rfi-hitachi-001.pdf.

– Data: quality of data and access to them are two essential variables. "AI applications are only as good as the data used to train them" [STU 19]. It is essential to respect people's rights when data are exploited.

Does the 2019 consultation express AI issues differently from the 2016 consultation? A comparison may be difficult, as the focus of the consultations is different (the 2019 consultation focuses on issues of norms and standards). However, using just one example, we can see that the way contributors approach AI issues is influenced by a variety of variables. For example, the way China is mentioned in the 2016 consultation is neutral and does not show any bias. China is then mentioned to remind us of the existence of a multi-year investment plan for robotics decided by Beijing. This is no doubt a way of calling on the American authorities to remind them to do the same. In the 2019 consultation, China is no longer a model. The look has changed, the expression is critical, negative. China, whose choices to use AI are portrayed in the West as the realization of an Orwellian society, now represents the antithesis of the ideal society, the counter-example, the construction that should be avoided. In the 2016 consultation, this threat of social control found its source rather in the American authorities themselves. The analyses and proposals for AI governance therefore use examples taken from the immediate news (in 2016, the United States was shocked by Edward Snowden's revelations; in 2019, the Western media described the new Chinese system of social control). But the basic idea remains the same: citizens must be vigilant about the practices of States, of their own governments. The two examples cited also show that vigilance is of little use in thwarting the authorities' plans.

3

Artificial Intelligence and Defense Issues

From the introduction of this book, we wanted to recall how AI came to be considered as a "cyber" technology, with AI a building block contributing to the expansion of the latter. AI is therefore an object that has found its proper place in the processes of appropriation of NICTs in the military field for several decades now.

3.1. Military policies and doctrines for AI: the American approach

3.1.1. *American defense AI policy*

3.1.1.1. *AI in speeches by the Secretary of the Department of Defense*

The online archive published on the United States Department of Defense website provides 1,310 transcribed speeches covering the period 1995–2019. However, not all the speeches delivered during this period are available. We therefore have only a subset, but the corpus is sufficiently complete to offer a vision of the evolution of the way in which problems, priorities and concepts are taken into account in the thinking of American defense officials.

Simply counting the terms used throughout the corpus, we have observed the position these concepts have been given (the figures provided here represent the number of times the term or expression occurs within the whole corpus).

The first observation is that so much attention is still paid to AI that all the speeches dealing with "cyber" issues are centered on the terms derived from it, and on highly generic terms (cyberspace, Internet, computer). The term "smart" also prevails over the term "intelligent" and the robot precedes AI. Although the frequency of terms alone cannot reflect the importance given to an object, the gap between the cyber domain, in the broadest sense, and the treatment of AI is such that the latter appears to be of lesser importance. For comparison, "nuclear" is mentioned some 2,126 times and "missiles" are mentioned 2,328 times. AI is relegated to the background of analyses issued by the Secretaries of Defense. This simple observation therefore goes against the image one might have of American defense strategies and priorities. Although, on the one hand, military investment in R&D, arms acquisitions or systems equipped with AI seems very substantial, AI has not yet acquired a priority status that would make it one of the central objects of discourse. More likely, it is understood, implied and implicitly integrated into the very broad "cyber" theme.

Term, expression	**n**
Cyber[1]	1,910
Internet	396
Compute[2]	394
Smart	255
Robot[3]	83
Artificial Intelligence	40
Intelligent	21
Machine Learning	10
Deep Learning	8
Autonomous weapon	5
Expert system	1
Neural Network	1

Table 3.1. *Presence of concepts in the corpus of speeches by the US Secretary of Defense*

1 "Cyber" includes here all the derivatives used such as *cyberspace*, *cybersecurity*, *cyber threat*, etc.

2 "Compute" and all its derivatives such as *computer*, *computing*, *computation*, etc.

3 "Robot", including its derivatives such as *robotics*.

3.1.1.2. *DARPA's strategic computing initiative (1983)*

On October 28, 1983, DARPA published "Strategic Computing. New-Generation Computing Technology: A Strategic Plan for its Development and Application to Critical Problems in Defense" [DAR 83]. By combining advances in AI, computing and microelectronics, the agency wants to create a new generation of technologies for intelligent machines. At this point, DARPA considered it appropriate to use the results obtained in various fields: in expert systems, in vision, in language (speech), in the understanding of natural language by machines, all of which now allow interaction between humans and machines. Machine intelligence, which is the objective of this program, is therefore more than AI, which is only a necessary component. This achievement implies the use of AI, expert systems (which the report sometimes associates with AI and sometimes dissociates from it), new theoretical contributions from computer science, progress in machine architecture (computers), the design of microsystems and microelectronics. In a pragmatic way, the military can then hope that this scientific and technological convergence will open up prospects in terms of new armaments, in particular, autonomous systems capable of reconnaissance and attack missions: "The possibilities are quite startling, and suggest that new generation computing could fundamentally change the nature of future conflicts."

From the point of view of DARPA experts, the progress made by all these disciplines is significant and marks a breakthrough. It is possible to imagine machines equipped with intelligence resembling that of a human, to envisage real, natural interactions between humans and machines, between machines and their environment; it is possible to seriously consider giving machines the ability to carry out complex tasks autonomously. These machines will be at the service of citizens, as well as of leaders who will now have intelligent computers at their side to act as assistant advisers. The aim is not to replace humans with machines, but to relieve humans, whether citizens or managers, of a large number of subordinate tasks or to facilitate their reasoning, so that they can concentrate on the essential, or the most important: "As a result the attention of human beings will increasingly be available to define objectives and to render judgments on the compelling aspects of the moment."

3.1.1.3. *The MAVEN project and the creation of the AWCFT (2017)*

The MAVEN AI project is a Department of Defense project, launched in 2017, to use ML to distinguish individuals and objects in thousands of hours of video footage recorded by military UAVs. The Google company that had contracted with the Department of Defense to contribute to this project had to back down following protests by thousands of people. However, these incidents do not challenge the project, which is being carried out by a team created in 2017 specifically, the Algorithmic Warfare Cross-Functional Team (AWCFT) [DSD 17], whose leadership will be entrusted to the Under Secretary of Defense for Intelligence. This project is being implemented to accelerate the adoption of AI in the armed forces.

At the official announcement of the launch of this project, the Department of Defense recalled that AI must be integrated by the defense forces (AI and Machine Learning). In this text, it talks about urgency and the need to move quickly: "I remain convinced that we must do much more and move much faster." The AWCFT's mission is to "accelerate the integration of Big Data and Machine Learning within the Department of Defense". The document specifies what the armies expect from AI:

– Speed is one of the goals of AI, one of its added values: "The AWCFT's objective is to turn the enormous volume of data available to DoD into actionable intelligence and insights at speed." The challenge is to learn how to use the colossal volumes of data collected, created, stored and available and make them useful to the functions of armies. The systems that should benefit from these advances are those of tactical UAVs and the processing of images and videos produced by these UAVs, the use of which is set to grow continuously over the next few years [SHA 18a]. These capabilities must meet the needs of ongoing operations in the context of the war against so-called Islamic State (Defeat-ISIS campaign [ALL 17]). The AWCFT must accelerate the implementation of Big Data and Machine Learning to use the vast amount of data that the forces are collecting. The first of these tasks is part of the war against Islamic State and must speed up or automate the processing, use and dissemination of UAV data to overcome the inability of humans to process these masses of data (humans are overwhelmed by these volumes of data), to produce useful intelligence efficiently (the machine's analytical power is greater than that of humans to find relevant information in images, in fact, in all types of data) and much faster, to strengthen the military decision-making process. The urgency of the project is justified by the threat from Islamic State and the ambition to defeat the organization as quickly as possible. The objective was therefore to deploy an AI algorithm before the end of 2017 in a theater of operations. The method and the project are therefore not to invest in long-term R&D, but rather to exploit the technologies and the state of the art which military officials consider useful for their immediate needs [FRE 17]. The military will take these tools from the private industry, which is the only one capable of responding to the defined needs. AI and ML appear to be the only reasonable, conceivable and effective options. The technologizing of armies throughout recent decades has constantly increased and improved sensors (cameras, drones, satellites, interception of communications, etc.), the capacities of network infrastructures (ever more data, calculations, flows, speed of data transmission on networks, etc.), the capacity of the network infrastructure (ever more data, calculations, flows, speed of data transmission on networks, etc.) and the capacity of the military to use these tools, to connect, inform and link weapon systems and to strengthen SRI systems. But today, this construction needs additional capacities, which cannot be found only in the speed and volume of flows. Intelligence and analysis has to go one step further. The crossing of this new threshold would, according to the US military, be indispensable for guaranteeing victories in the many armed conflicts underway and to come, just as networks, computerized systems and the domination of information space and cyberspace had to guarantee them before that. It could even be said that the armed forces and the intelligence agencies, and even administrations, have developed and learned to exist, decide, act and evolve in an environment that is totally overwhelmed by masses of data over the past several decades. The arrival of information technology marks this new impetus in the production of data. Machines process the data inserted into them, as well as producing new data in addition to the existing ones, in a process of exponential growth. Networks and the computerization of the planet have accelerated this phenomenon. This situation of data overabundance is not unique to the 2010s. Neither, therefore, is its capacity of overwhelming human beings, who are incapable of absorbing, or even reading, all the reports produced by humans and machines. We must therefore question the assertion of this seemingly sudden urgency as a vital necessity. It should also be recalled that the obsession with mastering the world through data, and therefore computers, has also been ingrained in the minds of the leaders, both military and political, for several decades. One will recall some accounts of situations during the Vietnam War, where the American military tried to rely on computers, asking them to predict the outcome of the conflict [MAD 17]. Strategists expected to read the future, not in crystal balls, but in quantitative analysis techniques, applying methods used in business to the business of war. This is reminiscent of the declarations of the American military who, in 2017, set up the MAVEN project, stating that it is sufficient to draw from the catalogue of industry methods to deal with, and settle, war affairs. The logic remains the same, faith in the computer, in computing power and in the power created by the data, but on the condition that it is massive, even belief in the effectiveness of a simple transposition of the logics and tools of commercial management to the management of war affairs. In the 1960s, during the Vietnam War, computers and programs were called the *Hamlett Evaluation System* (HES), and produced 90,000 pages of reports per day, impressive volumes of data, of which we will never really know what percentage was read, exploited and truly useful. The constant rationale consists of making inseparable informational domination, information control, military action driven by the results of data exploitation and victory. The entire American military apparatus was designed to produce phenomenal quantities of data. The strategy is the pursuit of victory through the production of large quantities of data, but this mass production is not synonymous with information dominance. The role of Robert McNamara, appointed as the Secretary of Defense in 1961, in introducing the use of computers as a decision support tool, is essential. But the defense forces, as well as state administrations, already have this culture of the calculator, of the exploitation of statistical data, which they have been using since World War II, and, for certain uses, since the 19th Century (use of punch card systems to process statistical data on the population) [HAR 88].

– The emphasis is on the problems of image and video processing (computer vision, videos) in intelligence functions (analysis, processing, understanding, etc.). The AWCFT must develop algorithms for processing this data and the identification and automatic detection of objects in the video images of UAVs, whose volume exceeds the processing capacities of human analysts and current systems. The algorithms to be developed by the AWCFT project will focus on image recognition, indexing and real-time detection of object classes in images and videos [SHA 18b]. The first algorithms had to be delivered by the end of 2017, which implied an accelerated development. It is likely, of course, that the algorithmic solutions proposed in this short time frame will not be totally new realizations and will build on previous results, but will be improvements of the already developed codes.

In its forced technological race, the U.S. military is mobilizing private sector companies and leaders in the processing and collection of global data. Google is thus meeting the needs of the defense sector by helping it to develop AI for drones [CON 18] (assistance in detecting objects in images collected by drones, surveillance technologies). This collaboration is part of the MAVEN project.

3.1.1.4. *Department of Defense strategy 2018*

The Department of Defense's 2018 strategy for AI announced 35 years after the *Strategic Computing Initiative* and published in 2019 [DOD 19] is a text that complements or extends the 2018 National Defense Strategy [DOD 18a]. It is not a real surprise because, the defense world has already taken an interest in AI and robotics in various forms, intelligent machines, which are the subject of plans, programs and projects.

The major powers have been conducting AI research for decades, usually under the close watchful eye of their armed forces, and have repeatedly reported on their plans for military applications since the 1970s. This strategy reaffirms the army's willingness to appropriate AI technologies to initiate a turning point that does not only concern the arsenal but heralds a more profound change. This integration of AI:

– will require training not only of individuals, but also of the learning machines that will need human expertise;

– will adopt commercial products, take over 80% of a commercial application and complete it with 20% modifications to adapt it to specific military needs. The Department of Defense calls on the private sector to work with them.

On June 27, 2018, the DoD created the Joint Artificial Intelligence Center (JAIC) [DSD 18]. The document, published by the Secretary of Defense, recalls the priority given to AI by defense, already stated in the National Defense Strategy (NDS) of 2018. AI is fundamental because it can "change society and […] the character of war". The aim of the project is to prepare the defense forces for the integration of AI. According to the Secretary of Defense, this is a matter of preserving and increasing the military advantage. The integration of AI and its applications in the military framework is an economic, bureaucratic (business reform) and strategic (appropriating AI and learning how to use it in combat) issue. The impact of AI must be assessed in the broadest possible way, as it will spread to all levels of the military institution, potentially producing its effects in the organization, command, the art of warfare and bureaucracy. This project is described as essential and urgent, as are all defense AI initiatives. The creation of the JAIC extends efforts to integrate AI into American defense.

3.1.1.5. *The American third offset strategy (TOS)*

"Offset strategies" refer to defense strategies aimed at strengthening military capabilities to counter specific threats. These strategies are essentially technocentric and are built on the basis of cutting-edge technological innovations, although more political or organizational considerations are, of course, at the heart of the choices made: "The critical innovation was to apply and combine these new systems and technologies with new strategic operational concepts" [HAG 14]:

– the first "offset" strategy, during Eisenhower's presidency, in the 1950s, was aimed at countering Soviet military capability. The strategy was to develop nuclear deterrence;

– the second "offset" strategy, in the 1970s, was aimed at countering Soviet quantitative superiority. It was the subject of a Long-Range Research and Development Planning Program;

– finally, the third strategy, initiated at the end of 2014 by the Secretary of Defense Chuck Hagel in a speech announcing the launch of a new program, the Defense Innovation Initiative [HAG 14]. This new strategy is intended to compensate for the erosion of American military superiority, which is partly due to the proliferation of denial of access (anti-access) and area denial (A2/AD) capabilities. US forces confronted with these capabilities have seen their freedom of action eroded. This problem emerged in the 1990s. This phase of reflection for the development of a third strategy was therefore officially initiated in 2014, but it builds on previous policies and strategies. The challenge to be met would be that of the technological gap with Russia and China which, over the last decade, have constantly invested in programs to modernize their armed forces. The United States must maintain its projection capabilities, Hagel said, or the world will become a more dangerous place for the United States. The new opportunities offered by AI are not explicitly mentioned in the 2014 speech, which only affirms the main principles.

The strategy is not yet complete, and foreign national strategies are already emerging that could counter the American project. Russia, which has once again become a major challenge for the United States, will contruct its own strategy to oppose that of America. The Russian strategy could include countering the third strategy using the principles of the first (i.e. strengthening the nuclear arsenal and threat, developing tactical and strategic nuclear weapons systems); the second part of this Russian strategy could consist of replicating American programs (in 2012, for example, Russia created the Advanced Research Foundation (ARF), which is working on projects similar to those of DARPA in the technical and logical fields, among which of course artificial intelligence plays an important role [KAS 17]).

3.1.1.6. *The quadrennial defense review (QDR)*

In this corpus, composed of six reports over the period 1997–2018, we have searched for the presence of terms dealing with cyberspace and objects related to AI and robotics, in order to situate them in time, to observe when they enter into defense policy considerations, and their respective weight in relation to each other. The results are presented in Figures 3.1 and 3.2.

It should be recalled that, in February 2018, the QDR was replaced by the National Defense Strategy, only a summary of which is made public.

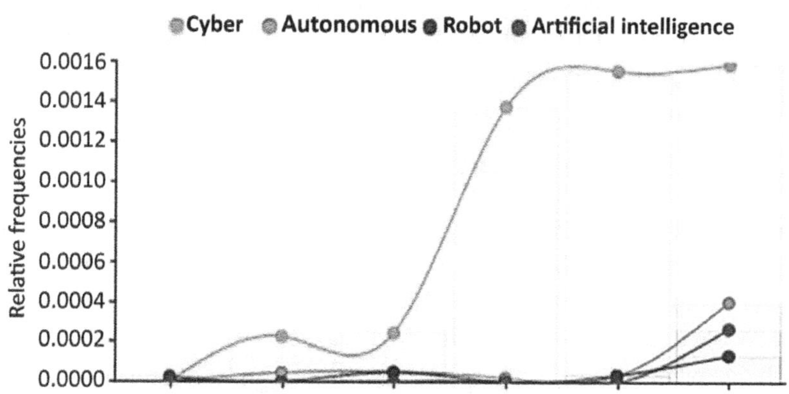

Figure 3.1. *QDR (1997, 2001, 2006, 2010, 2014) and NDS 2018. For a color version of this figure, see www.iste.co.uk/ventre/artificial.zip*

While the subject "cyber" has been well-considered in US defense policies as of 2001, AI does not appear in these documents until 2018. Robotics and autonomy are also subjects that are still marginal compared to the issues of cyberspace. All these topics are absent from the 1997 QDR. Overall, therefore, all these topics are less than 10 years old in these strategic documents.

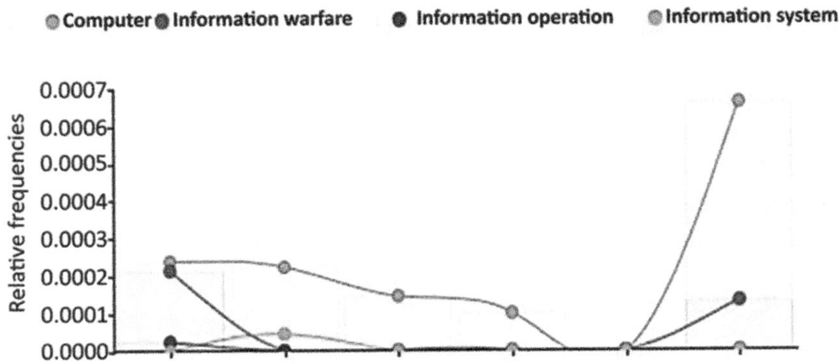

Figure 3.2. *QDRs from 1997 to 2018. For a color version of this figure, see www.iste.co.uk/ventre/artificial.zip*

However, themes that were present in 1997 around the issues of information warfare, information operations, information systems, information technology (computer, computing), and had since given way to emerging cyber themes, reappear in the latest document, from 2018. It should also be noted that there is no reference to the concept of "expert systems" in any of these documents.

3.1.2. *AI in American military doctrines*

Although AI has been in the research and development phase since the 1960s, on the initiative of the military (in particular, via its research agency, DARPA), if applications have been implemented, it is interesting to analyze the place occupied by AI in military doctrine. To do this, we have gone through a very large corpus of doctrinal texts (composed of Joint Publications), looking for any reference to AI, whether explicitly by the use of the terms "artificial intelligence", or by reference to "expert systems", "Machine Learning", "Deep Learning" and "autonomous". We have limited our search to the terms that seem to us to be the most representative of AI in recent years and currently (the table listing all the documents that we analyzed is provided in Appendix 2 of this book).

References to AI are therefore rare in the entire doctrinal corpus of the Department of Defense (here limited to Joint Publications):

– In 2017, in "Joint and National Intelligence Support to Military Operations. JP 2-01" [JCS 17], AI is essentially a tool to reduce the time taken for massive data processing. AI is an accelerator of processes linked to data processing:

> "Advances in data processing, such as artificial intelligence, large data-set analytics, knowledge bases, and iterative search tools, have created a new paradigm in which the timelines of intelligence operations and the intelligence process are greatly compressed."

– In 2019, AI is briefly discussed in "Joint Logistics. JP 4-0":

> "The joint logistics community must focus on the following five areas to influence mission success: warfighting readiness, competition below armed conflict, global integration, innovation, and the strengthening of alliances and partner networks […] A 'data culture' improves understanding of potential concepts like big data, artificial intelligence, machine learning, and modern computing power with regard to revolutionary improvements across the JLEnt. Adversaries will focus efforts on eroding the comparative competitive advantage in technology. Success in future conflicts may depend on the ability to expeditiously adopt and field new technologies that assure the continued ability to project and sustain power." [JCS 19]

– In 2016, in "Barriers, Obstacles, and Mine Warfare for Joint Operations. JP 3-15" [JCS 16], there was talk about autonomous systems:

> "Unmanned Systems PLT. The UUV team uses the Mk 18 Mod 1 UUV to search for and map underwater objects […] Mk 18 Mod 1 can operate for over 20 hours […] It is programmed using a laptop computer, and can employ sound-emitting transponders as navigational reference beacons, or its onboard computer can autonomously select another more appropriate navigation method to use. […]The unmanned aerial vehicle team uses the radio-controlled, man-portable *Silver Fox* and *Manta* unmanned aerial vehicles […] The aircraft are capable of

> fully autonomous flight and are able to carry various payloads to enhance mission effectiveness and adaptability."

– In 2014, the document "Joint Airspace Control. JP 3-52" [JCS 14] dealt with drone autonomy:

> "Unmanned Aircraft. […] UAS communication links are generally more critical than those required for manned systems. […] Although some UA may be capable of autonomous reaction (i.e. collision avoidance) or preplanned response in the event of lost communication (i.e. return to base), UA typically rely on a near-continuous data exchange for both flight control and payload management."

This work on the corpus of military texts can be completed by reading all the directives and instructions of the Department of Defense (a list of which is available on the fas.org[4] website). We note, in particular, the following texts:

– "Autonomy in Weapon Systems", November 21, 2012, "Incorporating Change 1", May 8, 2017. This Department of Defense directive [DOD 12], defines the conditions for the development of autonomous and semi-autonomous functions in weapon systems that apply lethal and non-lethal, kinetic and non-kinetic force. Not all weapon systems are affected by the rules on autonomy in this directive, for example, autonomous and semi-autonomous systems for operations in cyberspace, and certain categories of weapons[5]. However, no reference to AI is made in the 15 pages of the text (nor to the following terms: smart, robot, intelligence, Machine Learning). A definition of an autonomous weapon system is provided in this document:

> "Autonomous weapon system. A weapon system that, once activated, can select and engage targets without further intervention by a human operator. This includes human-supervised autonomous weapon systems that are designed to allow human operators to override operation of the weapon system, but can select and engage targets without further human input after activation."

4 https://fas.org/irp/doddir/dod/index.html.

5 Page 2 of the document.

The directive is a reminder of the legal and ethical framework that must govern the use of such autonomous systems:

> "Persons who authorize the use of, direct the use of, or operate autonomous and semiautonomous weapon systems must do so with appropriate care and in accordance with the law of war, applicable treaties, weapon system safety rules, and applicable rules of engagement (ROE)."

AI is only rarely invited into the multitude of themes and subjects covered by all these documents, which reflect the policies, strategies and debates that are taking place in the field of defense. Finally, let us analyze the documents produced by the various armies: of the land, of the air and of the sea.

3.1.2.1. *U.S. Army and AI*

For the Army, none of the documents listed[6] have a term in their title relating to intelligence or robotics or autonomy:

– However, robotics and autonomy are discussed very briefly in "Cyberspace Operations Concept Capability Plan 2016–2028", published in 2010:

> "Developments in quantum computing and nanotechnology may lead to a fighting force enhanced by robotics and remotely guided, autonomous, and miniaturized weapons systems. Communications systems may be self-organizing and distributed." [USA 10]

– In "Cyberspace and Electronic Warfare Operations 2025–2040", published in 2018 [USA 18], AI, robotics and autonomous systems are also mentioned:

> "The increased use of autonomous devices on the battlefield, including unmanned aerial systems provides challenges to security. The enhanced development of autonomous technologies portends a future where machines make decisions for themselves on the battlefield using advanced algorithms and artificial intelligence. Consequently, that decision-making algorithm may be hijacked and the artificial intelligence

6 We rely on the documents published on https://fas.org.

> corrupted, posing a danger to Army forces and technologies. Because of the proliferation of autonomous systems, fail-safe technologies and software are required to maintain positive control. [...]
>
> Autonomous active cyber defense. Develop systems which provide the ability for autonomous network defense through a collection of synchronized, real-time capabilities to discover, define, analyze and mitigate cyber threats and vulnerabilities without direct human intervention. This capability includes sensor-based artificial intelligence that learns and manages network topologies. [...]
>
> The Army employs more expeditionary cyberspace operations capabilities reducing the size, weight, and power of many cyberspace and EMS systems to make them more suitable for employment by remote, robotics, and autonomous systems. The Army deploys advanced antenna technology and dynamic spectrum access capabilities to increase efficient use of the available EMS."

– "The U.S. Army in Multi-Domain Operations 2028", published in 2018 [TRA 18a], also cannot ignore the entry of AI, along with other technologies, into military affairs. The document formulates generalities of a political or strategic nature:

> "emerging technologies like artificial intelligence, hypersonics, machine learning, nanotechnology, and robotics are driving a fundamental change in the character of war. As these technologies mature and their military applications become more clear, the impacts have the potential to revolutionize battlefields unlike anything since the integration of machine guns, tanks, and aviation which began the era of combined arms warfare [...] Unlike Russia, China has the economy and technological base, such as an independent microelectronics industry and world-leading artificial intelligence development process, sufficient to overtake current Russian system overmatch in the next 10–15 years."
>
> "The key to converging capabilities across all domains, the EMS, and the information environment is high-volume analytical

> capability and sensor-to-shooter links enabled by artificial intelligence, which complicates enemy deception and obscuration through automatic cross-cueing and target recognition."

AI will therefore be integrated as an accelerator (in the processing of masses of data) and a multiplier of military intelligence capabilities (enabling more data from multiple sources to be cross-referenced, in ever shorter timescales and with greater accuracy and quality). This AI contribution to the intelligence function is also reflected in "Intelligence. 2020–2040", published in 2017 [TRA 17], where it should reinforce the quality of the collections, the treatments, the speed, the precision and *in fine* fluidify the military action:

> "integrate artificial intelligence to enable autonomous threat detection and tracking [...] Artificial intelligence and autonomous, fast flight for small unmanned aerial systems will improve teaming between personnel and autonomous systems for highly mobile reconnaissance [...] Real-time event processing, artificial intelligence, and neuromorphic computing enable machine learning to enable more accurate and faster all-source understanding of complex situations."

We will remember from these developments that the military believes that, in AI, it will find, above all, an instrument to lift, at least partially, the fog of war and reduce friction. It is in this field that AI seems to be able to strengthen armies and in this field that international competition between armies engaged in a race for technology is played out. AI is not mentioned here as a new category of weapon, giving a more powerful or particular strike force.

In the "U.S. Army Concept for Multi-domain Combined Arms Operations at Echelons Above Brigade. 2025–2045", published in 2018 [TRA 18b], AI mainly influences two variables: human resources (improved performance) and decision-making processes (accelerated and improved): "Information technology and artificial intelligence (AI) may enhance staff productivity, lower staff personnel requirements, and speed and improve decision making in the future."

In October 2018, a task force dedicated to AI was created within the U.S. Army: the Army-AI Task Force (A-AI TF). This creation is announced in the Army Directive 2018-18 (Army Artificial Intelligence Task Force in

Support of the Department of Defense Joint Artificial Intelligence Center) [ESP 18]. What is at stake is the appropriation of AI by the U.S. Army, in line with the new strategy of the Department of Defense. This task force will be the link with the Joint Artificial Intelligence Center (JAIC) and will help define an AI strategy for the U.S. Army as quickly as possible. The task force will be located at Carnegie Mellon's National Robotics Engineering Center (NREC)[7] in Pittsburgh (a robotics research center working closely with the various actors of the U.S. Defense since 1995).

The 2018 directive lists the few topics the task force will focus on: Machine Learning, Deep Learning, Automation, Autonomous Targeting, Vehicles, Long-Range Precision Fires, Predictive Maintenance (to reduce costs and increase availability), Data Science, Hardware AI, Cloud Security and Accessibility, and Cybersecurity Issues for Integrating New AI Applications.

3.1.2.2. *U.S. navy and AI*

– In "OPNAV INSTRUCTION 5513.1E. Department of the Navy Security Classification Guides, 2005" [DON 05] in 2005, AI is taken into account thus:

> "Computer Resources. [...] Care must be taken to separate militarily sensitive information and database domains from non-military applications and/or architecture. Such categories would include:
>
> - System. Including applications, languages, tools, methodologies, management, artificial intelligence;
>
> - Artificial Intelligence. Including knowledge-based (expert) systems, robotics, image processing, natural language processing, speech processing, neural networks."

3.1.2.2.1. The NCARAI

The Navy Center for Applied Research in Artificial Intelligence (NCARAI)[8], a research center founded in 1981, is a component of the Information Technology Division within the Naval Research Laboratory. An

7 https://www.nrec.ri.cmu.edu/nrec/solutions/defense/index.html.

8 https://www.nrl.navy.mil/itd/aic/.

Autonomous Systems Laboratory was specifically created within the Naval Research Laboratory in 2012.

At the *Symposium on AI Applications in the Intelligence Community* in October 1983 [ECK 84], the center presented its research themes, which, at the time, were focused on the following areas:

– expert systems in combat management;

– expert systems for equipment troubleshooting;

– target classification, using ISAR data on vessels;

– natural language processing, applied to Navy message automation;

– multi-sensor fusion;

– adaptive control.

The BATTLE system was also presented at the symposium, the aim of which was to provide weapon allocation plans to the Marine Corps Artillery and Air Support. This project was part of a combat management program. The idea was to merge, in real time, the data brought up from the field by several observers on the damage caused to targets.

NCARAI has been working, since its inception, to develop applications for the U.S. Navy. The center is now organized around four main themes[9]: Intelligent Systems (cognitive science, cognitive robotics, human–robot interaction), Adaptive Systems (ML, autonomous systems, mobile robotics), Interactive Systems (developing interfaces for autonomous and intelligent systems) and Perceptual Systems (active and passive sensors for autonomous systems).

3.1.2.2.2. NCARAI publications

The NCARAI corpus of publications[10] offers 337 titles (online), covering the period 1990–2018. These publications are divided between technical (information technology) articles and articles combining cognitive sciences, human and social sciences, and operational and strategic considerations. For example, researchers are interested in the "behaviors" of intelligent

9 https://www.nrl.navy.mil/itd/aic/research.

10 https://www.nrl.navy.mil/itd/aic/biblio.

agents in works on the theme of AI revolt or rebellion [AHA 17, COM 17]: while human behavior such as protest, contestation, objection, saying "no" or rebellion can be beneficial and contribute to maintaining the balance between individuals, is necessary for individuals and societies, justified by imperatives of social justice, and is ethically, morally and politically acceptable, the same is not true of AIs, which are generally feared or interpreted in a negative way, and any possibility of rebellion is perceived as dangerous. This work considers AI, a set of artificial agents that play a social role, in its relationships with humankind, and considers AI rebellion no longer as a threat or destructive, but rather as potentially useful to these interactions. For, as with individuals, "rebellion" may not only be useful or beneficial, but also ethically or morally obligatory, or necessary. A constructive form of AI rebellion must be considered, just as it can be for human beings, in order to maximize the societal benefits of AI. AI rebel agents are defined as AI agents capable of developing attitudes of opposition to goals or tasks assigned to them by other agents. The "rebellion" can then take several forms: resistance, objection, the refusal to perform a task. An agent against whom an AI rebels is called an "interactor", and the latter is in a power relationship with the rebel agent. The "interactor" can be human or synthetic/artificial, an individual or a group. The authors then describe a typology of rebellious agents (agents who are aware of their rebellion, agents who are unconscious of their rebellion, those who are both aware of and capable of reasoning about the consequences of their rebellion – rebellion by design agents), interactors, types of rebellions, motives for rebellion and stages of rebellion. By cross-referencing these multiple variables, the authors draw two scenarios of AI agent rebellion (the first considers the participation of an intelligent agent in a disaster relief mission, in which intelligent agents refuse to execute orders received because they detect an error in the instructions; the second scenario considers a co-creation situation), but the combinations are practically infinite on the basis of the proposed typologies. Other works deal with autonomy, the place of error in reasoning, the detection of deception, the role of deception, trust, human–robot interaction, informational uncertainty, etc.

A statistical analysis of this corpus of publications allows us to observe some trends in the consideration of topics such as robotics, expert systems or autonomy (Figure 3.3).

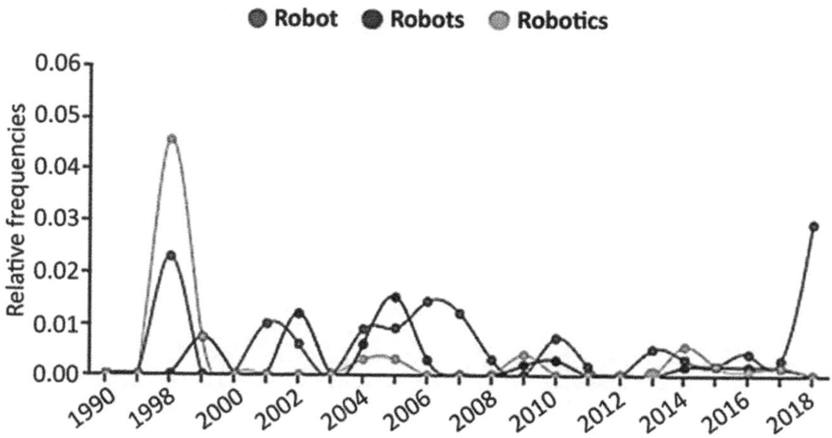

Figure 3.3. *The theme "robot" (robot, robotics) in NCARAI publications. Period 1990–2018. Data processed using the Voyant Tools application. For a color version of this figure, see www.iste.co.uk/ventre/artificial.zip*

While robotics has been treated relatively consistently over the period of almost 30 years, the issue of autonomy has also been taken into account since the 1990s, but on a more regular basis as of 2010.

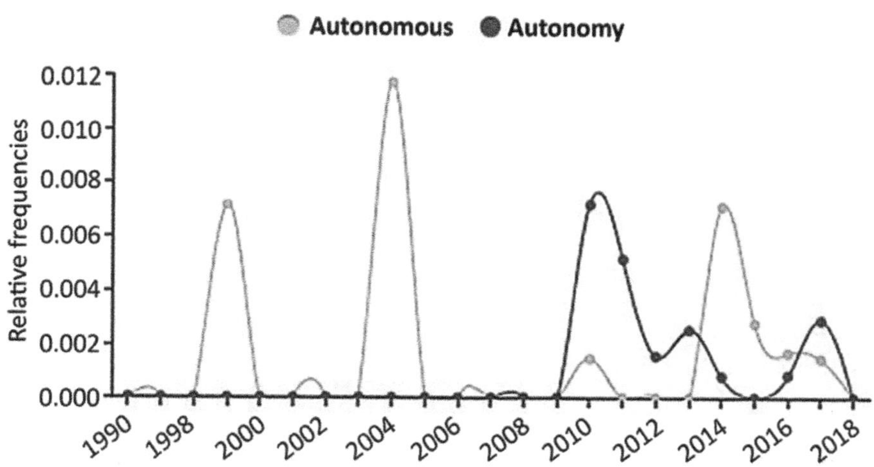

Figure 3.4. *The theme "autonomy" (autonomy, autonomous) in NCARAI publications. Period 1990–2018. Data processed using the Voyant Tools application. For a color version of this figure, see www.iste.co.uk/ventre/artificial.zip*

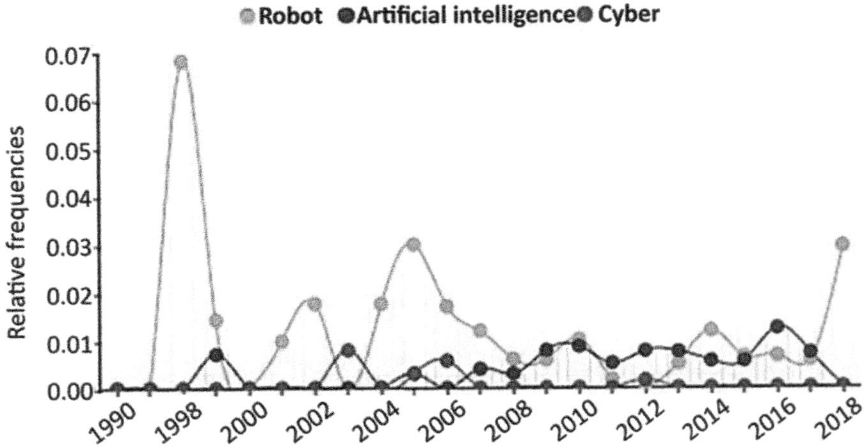

Figure 3.5. *The "robot" (robotics), "artificial intelligence" and "cyber" themes in NCARAI publications. Period 1990–2018. Data processed using the Voyant Tools application. For a color version of this figure, see www.iste.co.uk/ventre/artificial.zip*

During the period under review, robotics-related issues were predominant, with considerations on cyberspace itself minor. AI has overtaken the "cyber" question, which is only treated, as such, on an *ad hoc* basis between 2002 and 2006.

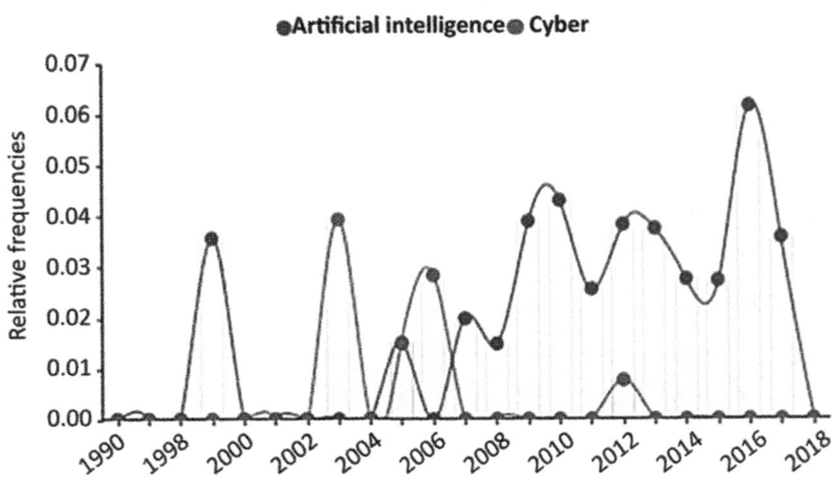

Figure 3.6. *The "artificial intelligence" and "cyber" themes in NCARAI publications. Period 1990–2018. Data processed using the Voyant Tools application. For a color version of this figure, see www.iste.co.uk/ventre/artificial.zip*

Finally, it should be noted that none of the articles introduces, at least in their titles, the notions of "expert system", "war" or "warfare", "crime", "security"; the notion of "defense" is practically ignored as well (it is used only once). The predominant concepts are those of "cognitive system", "human", "society", "reasoning", "learning", "robot", "interaction", "planning" and "autonomy".

3.1.2.3. *U.S. Marine Corps and AI*

In the "MCDP 6 Command and Control", published in 1996 [DON 96], expert systems are mentioned, but the notion of AI is not mentioned:

> "Expert systems and artificial intelligence can assist with cognition to a certain extent – by helping to integrate pieces of processed data, for example. But cognition is primarily a human mental activity – not primarily a procedural act like processing, but an act of learning.
>
> We transform the complex components of knowledge into understanding through *judgment*, a purely human skill based on experience and intuition, beyond the capability of any current artificial intelligence or expert system. Judgment simply cannot be reduced to procedures or rules (no matter how complex)."

In 2003, "MCWP 3-42.1. Unmanned Aerial Vehicle Operations, 14 August 2003" [USM 03] deals more specifically with the use of UAVs:

> "unmanned aerial vehicle – A powered, aerial vehicle that does not carry a human operator, uses aerodynamic forces to provide vehicle lift, can fly autonomously or be piloted remotely, can be expendable or recoverable, and can carry a lethal or nonlethal payload."

3.1.2.4. *U.S. Air Force and AI*

The U.S. Air Force doctrine over the years has dealt with autonomy, AI and expert systems, but no formal doctrine document appears to address AI in its entirety or as the sole focus of its purpose.

For the notion of "autonomy", we have referenced four documents for the period 2008–2017:

– "AFDD 3-01, Counterair Operations, 1 October 2008, interim change 2, 1 November 2011": "To prevent fratricide, great caution should be exercised when employing autonomous combat identification in defensive counterair operations";

– "AFDD 3-40, Counter-Chemical, Biological, Radiological and Nuclear Operations, 26 January 2007, Interim Change 2, 1 November 2011": "unmanned aircraft. An aircraft or balloon that does not carry a human operator and is capable of flight under remote control or autonomous programming";

– "Energy Horizons: United States Air Force Energy S&T Vision 2011–2026, AF/ST TR 11-01, 31 January 2012" [SAF 12]: "future RPAs and autonomous aircraft could be tailored to specific mobility and combat missions currently accomplished by traditional aircraft, and do so with a reduced total energy footprint"; "autonomous operation of bird-sized micro air vehicles demand high capacity computer operations be carried out in physical spaces equivalent to golf ball sized brains";

– "Air Force Instruction 63-101/20-101, May 9, 2017" [SAF 17]: "When developing autonomous and semi-autonomous weapon systems, assess the requirements and guidelines in the directive (DoDD 3000.09)."

In the same way, for the concept of AI, we have identified three documents for the period 2005–2017:

– "AFDD 3-13, Information Operations, January 11, 2005, incorporating Change 1, July 28, 2011", simply refers, in its bibliography, without using the notion of AI in the document, to the publication "The Cult of Information: A Neo-Luddite Treatise on High-Tech, Artificial Intelligence, and the True Art of Thinking"[11];

– "Energy Horizons: United States Air Force Energy S&T Vision 2011-2026, AF/ST TR 11-01, 31 January 2012" [SAF 12]:

> "The AF vision for the more distant future involves redundant, fractionated, cooperative systems that operate in contested, congested, and competitive environments. To be realized, this vision requires fundamental new ideas in artificial intelligence

11 Roszak T., *The Cult of Information: A Neo-Luddite Treatise on High-Tech, Artificial Intelligence, and the True Art of Thinking*. 2nd edition, University of California Press, Berkeley, 1994.

as well as control theory that go well beyond today's UAV and robotics research."

– "Air Force Manual 15-129, Vol. 2 (2011/2017)" [SAF 13]:

"Integrated Weather Effects Decision Aids (IWEDA). IWEDA is a rules-based TDA application. IWEDA supports both Army and Air Force systems, and improves interoperability. IWEDA uses artificial intelligence techniques and knowledge of atmospheric effects with model data to enhance and expand current weather decision capabilities. It allows commanders to compare weather-based advantages/disadvantages of friendly and enemy systems."

Finally, for the notion of "expert systems", two documents (from 2002 and 2011) are identified that mention it:

– "AFDD 6-0, Command and Control, June 1 2007, incorporating Change 1, July 28, 2011":

"C2 processes are the structured basis of informed decision-making. Technology either automates or accelerates these processes via advances in information technology like digital electronic communications, computers, and expert systems";

– Air Force Handbook, Vol. 5 36-2235, 2002 [SAF 02]:

- "An expert system recognizes mistakes commonly made by students, and can detect and diagnose errors, and present information to correct misconceptions that would normally require an instructor";

- "Because of the expense involved in development of expert systems, they are usually reserved for teaching high-risk knowledge and decision-making skills. Such training may be an appropriate and cost-effective way to prepare military personnel for major military exercises."

AI and its associated concepts are virtually absent from the corpus analyzed in the Air Force's considerations on issues related to information, cyberspace, information systems, data and information control. The U.S. Air Force has been building its doctrine and its entire development over the past few decades by appropriating information and communication technologies, asserting itself as a central player in the domination of information and cyberspace, while excluding AI from its discourse.

3.2. Military AI in Russia

The military in Russia, and throughout the world, believe that the next wars will be fought at a frantic pace and that under these conditions, men will no longer be able to keep up and decision-makers will no longer be able to receive information, process it and make decisions. The tempo will be accelerated. Only robots and intelligent, autonomous systems will be able to act in this environment. But the acceleration is also due to the introduction of these systems. Defense expert Vasily Sychev believes that the systems themselves do not pose a threat to humankind, because machines will not revolt against humans. If intelligent, autonomous and increasingly powerful systems of destruction are a threat, it is because of their owners' sense of impunity [SYC 15].

Vadim Kozyulin [KOZ 18] identifies three threat groups emanating from autonomous weapon systems driven by AIs:

– risk associated with excluding humans from the decision-making loop;

– risk of undermining international balances, by allowing non-nuclearized states to play on an equal footing thanks to new-generation weapons based on AI (AI as a deterrent weapon because of its aggressive and defensive potential);

– risk associated with reducing the time required for decision-making and response. Computers and networks have made it possible to design C4ISR systems that exceed the capabilities of human thinking, decision-making and response. The analysis of masses of data is now the task of supercomputers, responsible for extracting information and producing scenarios for political and military leaders. However, the latter are caught in a tempo that is accelerating and no longer allows time for decision-making, which, in any case, is based on data from algorithmic processing.

According to him:

– with AI, everything that can be automated and robotized will be;

– the traditional armed forces will be reconfigured and combined to produce a more effective action;

– the exchange of information along the horizontal and vertical axes (e.g. ground-to-air, from tank or infantryman to airplane pilot) will be optimized. Not all players will receive all information, but only that which is useful to them, that which the machine decides is relevant (according to a multitude of tactical, operational and strategic criteria).

Sergei Abramov, director of armament industries, believes that fully autonomous weapon systems cannot be implemented without prior certainty as to the nature of the target. Autonomous weapon systems based on AI already exist, but there can be no question of replacing humans with machines. He argues that the purpose and utility of the autonomous weapon is not to replace humans, but rather to give humans greater control over systems, what they are asked to do, and how they do it [MAK 19]. The serious incident in South Africa in 2007, in which nine soldiers died, is cited as an example of the potential flaws in robotic or automated systems (although experts have subsequently dismissed this explanation): in October 2007, during military exercises, an out-of-control Oerlikon anti-aircraft gun killed and wounded several soldiers. The gun is equipped with an automatic target detection and tracking system, but remains under the control of the human operator. The incident has become, rightly or wrongly, an illustration of the risks that out-of-control robots run when their systems are faulty (bugs, attacks).

Russian literature on the militarization of AI and its place in modern wars and defense affairs, formulates arguments that are generally favorable to the development of these new technologies, considering the process, on the one hand, as inevitable, and, on the other hand, as necessary to face difficult theaters of operations (urban wars) and to compete on an equal footing in terms of armaments with the great competing powers of the planet. There seems to be no alternative to AI, so it is no real choice. The technology exists and it must be used. Regulation, legislation and possible international agreements to regulate their use could be envisaged at a later date. However, reflections on the place of AI in the art of warfare do not, rule out the question of the relationship between humans and the intelligent machine, on the contrary, considering it essential to maintain humans in command functions and strategic decision-making centers, while recognizing that they are in a state of inferiority in relation to these machines, unable to process data in the same volumes and at the same speed as the systems. AI is thus vital in the attempt to master cyberspace which, at the heart of its operation, has already sidelined humans. None of the articles that we have been able to read envisages any other configuration than a continuation of the mastery of the art of war by increasing data and transforming it into reasoning, decisions and actions. War can no longer be fought without massive data. AI is indispensable to give meaning to these data.

Here, we offer an overview of some ideas developed in the recent Russian literature on AI in war. These reflections are developed on Russian websites specialized in defense or political issues, for the most part. These ideas do not reflect the doctrine and official policy of the Russian authorities. However, they nevertheless reflect a form of Russian thought on these issues.

Wars and armies are becoming more and more technologically advanced

AI contributes to this inevitable phenomenon by making weapons, weapon systems, decision-making and data analysis processes more effective. This process cannot and must not be resisted:

– it is impossible to contain the development of militarized AI, because the technology is accessible, because wars are increasingly technologized and because AI will be useful in this process. AI can be used in many different types of applications [LOS 17]. AI, soldier–machine interaction, unmanned military vehicles, robots, autonomous weapons, directed energy and hypersonic weapons are all critical areas for modern weaponry. Automatic pattern recognition (APR) technologies allow for more rapid and efficient analysis of satellite images and radars; AI will increase the effectiveness of radar systems for attack detection ballistics; the multiplication of miniaturized satellites and the simultaneous growth in their observation capabilities call for fast, powerful data processing systems capable, in particular, of distinguishing military areas or objects (e.g. vehicles) in order to differentiate them from those used by the civil military; on the battlefield, AI will increase capabilities for visualization and, for example, recognition of the types of weapons used by the military enemies;

– the United States, in its hegemonic project, will sooner or later wage a military war, and not just an economic war, against Russia or China. For Russia, it is essential to pursue developments in breakthrough technologies, such as AI, to maintain strategic parity with the United States and China. It is by maintaining a level of technological and military power that Russia will guarantee its place on the international stage in a world where balances are fragile. In the race for defense technologies, Russia has lower financial capacities than the United States and China, and in this situation, AI could help to compensate for the imbalance. In particular, the expected gain would largely relate to the contribution of AI to strategic decision-making processes [LOS 17];

– the quest for power through AI has inevitably led to an arms race, which can be joined by non-state actors. The technological opportunities for terrorism are mentioned [LOS 17].

AI and robotics will give rise to new weapons but probably not androids

Although the long-term possibility of intelligent android robots cannot be ruled out, in the near future, it is more likely that drones, "smart" robots taking the form of conventional weapons, vehicles, industrial "machines", etc., will be developed. AI will be embedded in pre-existing armaments, whose operating modes it will increase or modify [LOS 17].

We cannot oppose the technological evolution that relies on AI because it is the only solution currently available to compensate for the inability of humans to process data, think and act at the same speed as technological systems.

The multiplication of the masses of data and the processing to be carried out, often in real time, and the mass of data to be interpreted are definitively beyond the reach of human capacities. The tasks to be performed by intelligent systems also include ensuring the cybersecurity of defense systems (detection of vulnerabilities, attacks, reaction, counter-attacks) [LOS 17].

AI can be applied to all military functions

There are four groups of tasks for militarized AI: informational, tactical, strategic and economic. Increased data collection capabilities combined with increased analytical capabilities will provide tactical and strategic advantages, based on the speed and quality of information processing [LOS 17].

AI has military applications that are not immediately lethal (e.g. improving logistics – known as intelligent logistics – increasing warning capabilities, attack detection, radar, etc.); applications in weapon systems and in the operations themselves, helping soldiers on operations; and applications at the level of decision-making processes (helping to define scenarios for strategic decision-making, for example). AI in military affairs can be defined as an application domain where systems, models and devices imitate human intellectual activity (perception, reasoning and logical decision) in military affairs [STE 19].

AI paves the way for new operational modes

The possibilities for using AI in tactical weapons are numerous. UAVs, which can now be produced in large quantities due to low costs, can be operated as fleets or swarms, managed by AI. For the same price as a fighter plane, an army could afford hundreds of thousands of drones which, equipped with explosives, would be formidable weapons. These technologies would make it possible to create "swarms" or "pack" weapons (missiles, UAVs, land robots, etc.), need AI, computerized and communicating systems, communication flows and systems that derive their performance from their adaptability and their capacities for analysis, collective decision-making and collective adaptation (such as the U.S. Navy's CARACaS application for managing swarms or fleets of robot ships and drone ships)[12]. Modern warfare strategies have required changes in the way troops are deployed. These new rules of deployment must integrate the consequences of the appearance of new generation weapons which also modify methods of fighting. Swarms of drones, the weapons to which tasks can be entrusted, which, until now, were only in the hands of soldiers and depended on human reasoning and decisions, require a review of the role of individuals and the interaction between humans and machines, humans and systems, AI and humans [LOS 17].

AI feeds the information space, constructs it and enriches it

AI enriches the information space by producing large amounts of artificial data to lure other systems, intelligent or not. These AI creations are

12 https://www.dsiac.org/resources/news/onr-developed-caracas-software-controls-navy-swarmboats.

called "virtual truth". These practices are not without risks, for those who are victims of these lures of virtual truth, whether it is political opposition groups, for example, who will be trapped, or governments, armed forces, etc., who will be trapped. The range of possibilities for manipulation will be increased [LOS 17].

America is portrayed as an aggressive player in the information field. Leonid Savin's article [SAV 19] refers to the practices of the American defense forces which aim to manipulate public opinion, of creating fakes and false traces in order to make accusations against third parties by using AI technologies, developed from the work of DARPA, which is said to have some 20 programs on AI and 60 on cybersecurity.

In 2018, the Minister and Deputy Defense Minister of Russia called on the scientific community and industry to join forces to advance in the field of AI, which is essential in cyberwarfare, and all the more central as no conflict today takes place without battles taking place in information space. Russia, they say, can build on its excellence in mathematical research[13]. With AI, a new era is opening up, they say, because the speed of calculation is no longer enough. Intelligent systems are needed to process and store information. All these technological contributions create a new generation of weapons, more powerful, faster, more precise, more resistant and undetectable.

AI is a key element of network-centric warfare [RUD 19]. The conflict against Georgia in 2008 was an opportunity to highlight several weaknesses in the Russian military organization: weak intelligence and communication capabilities, low degree of automation of the processes of controlling troops and weapons, weakness of the troops' information system. Faced with the volume of information produced by the systems, decision-makers are no longer able to make the right analyses and choices. The task is immense and far beyond human capacity in terms of complexity.

Only software solutions known as intelligent software solutions, based on neural networks, can handle the masses of data and complexity.

13 https://roskomsvoboda.org/37087/.

It is essential to take into consideration the question of the relationship between humans and AI

In modern, automated weapon systems, the place of humans remains indispensable. But its main weakness, compared to systems and AI, is speed. Humans are not capable of acting, deciding, reacting, analyzing and taking into account all environmental data at the same speed as the systems. A human being is, compared to these military technologies, fragile and inefficient, one of the slowest links, if not the slowest, in the decision-making chain [LOS 17].

The Deputy Defense Minister of the Russian Federation, Ruslan Tsalikov, said that the Russian armed forces have long been using AI. In the 1970s and 1980s, the fledgling AI had limited applications in the military field. In 2020, the situation has radically changed: there are autonomous drones and electronic assistants based on AI. AI realizes the dream of having soldiers who are impartial, precise, fast in the execution of their tasks, who clearly carry out orders, know how to quickly analyze a situation, are the first to fire (optimally managed OODA loop). Wars are always more technological and take unsuspected, sometimes invisible, forms. Tasks carried out by civilians may very well be part of the process of military operations without them being aware of it [NIK 19].

AI alone is nothing. It acts on the initial decision of humans, according to projects defined by them. It cannot become dangerous or violent without prior human intervention. Thus, we cannot expect machines to take command of armies, to make decisions in war independently. In the short term, there will be no C2 replaced by an AI. On the other hand, it is possible to implement unified defense systems, such as air defense, where the most appropriate means of destruction are compared to the targets to be reached, taking into account variables such as the level of threat they may represent [NIK 19].

Dependence on and effectiveness of intelligent systems does not preclude the issue of trust

Are AI-based decision-making systems infallible? For example, wouldn't an AI-equipped tank, which would have to make firing decisions in place of the tank commander, risk opening fire on its own soldiers or paratroopers, for example? The insurance lies in the knowledge that AI has. It must have all the knowledge (which it has to be taught) to distinguish enemy targets, to recognize its own soldiers and everything that is not an enemy. The lives and the safety of the soldiers then depend on the quality of the system and AI. The effectiveness of the systems themselves depends on their own security, their own inviolability (protection against hacking of AI in particular). AI is vulnerable to hacking, sabotage and means of disrupting its operations [LOS 17].

Can AI be a weapon of dissuasion? What is the strategic role of AI?

The strategic power of AI raises questions. Some believe that the mere presence of AI in weapon systems will eventually be sufficient to act as a deterrent, in the same way as nuclear weapons do [LOS 17].

AI could also be the cause of conflicts not decided by humans [LOS 17].

AI can also be an instrument of power within States, an instrument of economic, informational, political–military superiority, etc. AI thus conditions many variables in the political, economic and social life of States [LOS 17].

No global system or solution likely to significantly modify the free balance of forces seems conceivable. AI is not yet capable of radically changing the strategic balance in the same way as nuclear weapons did. AI is thus more a supporting, complementary, supporting technology, a force multiplier, but not a technological and strategic breakthrough [NIK 19].

The AI that currently exists cannot really be considered a form of intelligence. We have tools that are artificial eyes, artificial ears and artificial noses, but not yet a brain. This limits the applications of current AI to autonomous actions at a tactical level, but these systems are not capable of thinking on a more global level [NIK 19].

The economic dimension is a key factor that legitimizes military AI research and the integration of AI into armies and warfare.

In November 2018, during a speech at the ERA defense technology base in Anapa (a city on the Black Sea coast), Vladimir Putin called on industrialists to develop intelligent, high-precision weapons and missiles that increase the capabilities of existing and future weapon systems. In 2017, the ERA Defense Technopole counted some 37 companies, 160 researchers and 47 ongoing R&D projects on various subjects such as robotics, bio-engineering, nanotechnology, information systems, AI, etc. The ERA Defense Technopole was the largest in the world in 2017. There is an economic reason for this objective. Leaders believe that what was then achievable with very expensive weapon systems can be achieved with much smaller systems using high-precision weapons and ammunition [GAL 18]. This policy is constrained. The results must be achieved as quickly as possible and be of the highest possible quality. The production chain for these weapon systems is very long. However, in the face of other nations engaged in similar developments, Russia has a duty to be effective. This race can be won by shortening production cycles. AI is not only played out in research laboratories or in test and prototype development centers. In the military field, production is a key phase of international competition.

At the moment, there are potentially many applications for AI in the military, but certainly more than is economically feasible. Choices will have to be made [NIK 19].

It is difficult to assess the actual levels of state investment in AI R&D and military R&D. The level of investment in AI and military R&D is difficult to assess. In the West, most research is civilian. In Russia, the situation is different: the bulk of research is military, and then transfers are made to the civilian sphere. The financial efforts are thus concentrated on defenses, whose initial mission is the development of a new generation of intelligent weapons by 2025 [NIK 19].

3.3. AI and the art of warfare

3.3.1. *Manuel de Landa: war in the age of intelligent machines*

In 1991, in *War in the Age of Intelligent Machines* [DEL 91], Manuel de Landa offered us a prospective vision of the armed conflict transformed by the irruption of intelligent machines.

From the very first lines, the text opens with a reference to "killer robots", now taken out of the science fiction universe to which they were confined. These killer robots, which certainly have neither the appearance nor the intelligence of a human being, have already entered military history. Prowler, for example, is a land vehicle equipped with a vision system that analyzes images to allow the machine to move in a hostile environment, distinguishing between enemy and friendly forces. All these capabilities were still in the future, in the R&D phase within defense, at the time when Manuel de Landa wrote his book. Robots do not yet have real autonomy. The decision to fire against an enemy is still under the control of a human, who supervises the machine. But already, there are two categories of AI and smart robots: on the one hand, AI that helps and assists the military, and, on the other hand, one that realizes, executes and acts. At the beginning of the 1990s, the transition from one to the other was visible in war games. Until then, humans made the decisions affecting troop movements, the computer calculating the effects of the attacks carried out. But new automatons like SAM and IVAN were taking over the role of humans and initiating decision-making. For Manuel de Landa, the contributions of AI would materialize at several levels:

– in the "traditional" applications of information technology in warfare: "radar systems, radio networks for Control, Command and Communications, navigation and guidance devices for missiles." AI would make these applications "smarter" in line with the progress achieved by AI;

– AI was to be integrated into defensive and offensive weapon systems. However, the author argued that human-like robots that would put soldiers back in the field of battle, or robots that would provide command and control functions and conduct operations, should be excluded;

– the introduction of AI was to have an impact on the military and the art of warfare. In particular, it was to promote decentralization. "AI research is evolving toward a model of control based on dispersed decision-making, a model that could be used to promote decentralization within the military."

Here is a summary of the main ideas and hypotheses put forward by Manuel de Landa about AI and robots in 1990:

– AI is not advanced enough to consider the creation of robot killers;

– the line separating "advisor, assistant" AI and "decisional" AI is being erased;

– it will be another two decades before AI will be able to create autonomous weapons;

– expert systems are the most successful implementation of AI to date (the three applications of AI that the Pentagon wants in "Strategic Computing" (1984) mobilize expert systems (battle management advisers, cockpit advisers, autonomous weapons));

– one of the obsessions of military commands is removing the human element from the battlefield, removing the soldier from the decision-making loop. In "Strategic Computing" (1984), the Pentagon put forward this project again (to create autonomous weapon systems capable of fighting alone);

– the needs of intelligence agencies have decided some AI research priorities. The NSA employs a large number of linguists and translators. The translation of foreign languages is one of the major functions in intelligence. The automation of these tasks would thus have been one of the priorities in the history of AI[14]. In fact, the first AI research project was a machine translation program funded by the Air Force. The lack of quick results condemned the project, which was stopped in 1966;

– the introduction of expert systems into command procedures has the effect of centralizing the command function and moving it away from the battlefield. The commander becomes a battlefield manager, he has to manage data flows and procedures:

> "The centralized implementation of a battle plan involves, as we saw, an enormous increase in the amount of information that must be processed by the upper echelons of a command system. In such circumstances, the task of a supreme commander is reduced to that of an information flow manager. This impersonal, detached approach to battle management closely resembles that of World War I commanders who directed their battles from behind the lines [...] World War II commanders such as Guderian, Patton and MacArthur returned to the battlefield, becoming directly involved in the implementation of tactical plans. Half a century later, expert systems technology is creating the conditions for a return to the World War I style of

14 Manuel de Landa explains that this enthusiasm for machine translation would be the consequence of the success of cryptology during the war. Any language could, it was thought at the time, be decoded in the same way as encrypted messages.

> command, reducing once again the function of generalship to that of a 'battle manager'."

The central hypothesis of Manuel de Landa's work is that if the military wants AI and wants to create autonomous intelligent machines that can kill, it is to remove the human being, the soldier, from the decision loop and the battlefield. But as the author points out, this objective has been contested and alternatives have been proposed, to use these machines for other purposes, in a different way: rather than removing humans, the objective could be, on the contrary, to associate humans and machines:

> "There are, however, alternative uses of those technologies that are easier to develop and that do not pit machines against humans, but rather aim at creating a synergistic whole out of humans and machines [...] In the 1960s as the military was sponsoring research with a view to get humans out of the loop, independent researchers like Doug Engelhart began to work in the opposite direction: to create an interface between humans and machines capable of assembling them into a synergistic whole. Such researchers called their concept the 'augmentation of man's intellect'."

The idea would be to integrate humans and machines, to amplify the human's intellectual capacities.

However, it can take a long time between the emergence of technologies and their integration into armies and warfare. The tactical integration of new technologies has always been a long process. There are distinct temporalities: that of political projects and strategies (plans, multi-year programs, etc.); that of the construction of military doctrinal thought, which must anticipate, consider options, consequences and effects on the art of warfare and on the institution itself; that of science and technology, research and industry development; and that of industry.

3.3.2. *AI announcing a new RMA?*

Some, anticipating radical transformations in military affairs under the impulse of AI, evoke a new Revolution in Military Affairs (RMA). This permanent search for new modalities of power has led armies to take an interest in AI for several decades. Armies and the art of warfare have been marked by major technological developments, and AI may be one of those major new technological transformations that could have a profound impact on the art of warfare. However, nothing is yet certain.

In September 1985, *Le Monde diplomatique* published an article by Dominique Desbois entitled "*Comment 'l'intelligence artificielle' conduirait la guerre*" ("How 'Artificial intelligence' would lead the war") [DES 85]. The year 1983 witnessed an acceleration in military investment in the United States. DARPA initiated a new project[15] to develop a new generation of strategic technologies, in line with Ronald Reagan's defense policy (the "'Star Wars' speech", March 1983). We were also in the midst of the Cold War, a nuclear arms race and the risk of confrontation between the two blocs. On the one hand, there was a need to strengthen the nuclear arsenal, and, on the other hand, to increase the capabilities of C2 systems. Defense budgets would explode in the coming years. The budget for the *Strategic Defense Initiative* (SDI, a project centered on the development of ballistic missile interceptor capabilities) rose from $99 million in 1984 to $3.8 billion in 1986. The time taken to commit forces was reduced, and C2 was more removed from tactical and operational levels than before: in this context, the army had to acquire automated information processing facilities, and was planning to process ever greater volumes of data and to automate decision-making. Speed was a key element of this strategy. In terms of science and technology, it would therefore be necessary to concentrate efforts on increasing computing capacity (new hardware, in particular), and on the software dimension, efforts would focus on AI (more particularly, expert systems) to integrate natural language comprehension, vision and reasoning into systems.

	Needs
Army	Autonomous vehicles for hostile environments
Aviation	Intelligent co-pilot crew assistance system capable of autonomy
Marine	Battlefield management system capable of processing unverified data to predict probable events; system providing learning-based action strategies to assist in decision-making

Table 3.2. *Expressed military requirements. Reconstituted table from the article by Dominique Desbois*

15 Strategic Computing, New Generation Computing Technology: A Strategic Plan for its Development and Application to Critical Problems in Defense. Defense Advanced Research Projects Agency, October 1983.

Dominique Desbois observed that the American project aimed to "converge scientific research towards precise military objectives".

The ambition was thus to create machines completely autonomous in intelligence/reconnaissance as much as in attack and defense, at long distance. The two main challenges then were the instantaneity and unpredictability of war: for DARPA, AI was a solution. With expert systems, it was possible to envisage replacing humans by machines in processes requiring very rapid decision-making. The decision cycle was so fast it exceeded human capabilities. Only computers were able to process information with the right speed. When it came to countering ballistic missile attacks, the speed of decisions was critical. But when it came to these decisions, when we talked about putting the human element out of the loop, it meant no longer putting the highest decision-makers, in the case of the United States, the President himself, in the loop. The idea of replacing the White House with a computer provoked strong reactions in Congress. Pentagon experts were therefore showing their blind confidence in the technology and its capabilities (precision, speed, autonomy, absence of error), a confidence that was based, in part, on promises made by the scientists themselves. But researchers were struggling to make their hopes and promises come true. Technocrats were victims of beliefs that had been propagated by researchers and industry, and of misleading words themselves. Dominique Desbois rightly mentioned misunderstandings, shifts in the meaning of concepts, false meanings and corruption of language. But no one was fooled. While funds were once again pouring in to finance research and development in AI, a community of American computer scientists (the ACM) reintroduced the debate on computer security, on the vulnerability of computer systems, in 1984. The issues that are still at the heart of cybersecurity today are: the vulnerability of critical systems, i.e. for air transport, health systems, military systems and nuclear defense systems. Computer scientists are aware of the limits of information technology, of the risks of errors that could lead to a nuclear confrontation.

DARPA's "Machine Intelligence" project (2011–2016) defines some of the challenges of AI for defense:

> "The Machine Intelligence project is budgeted in the Applied Research Budget Activity because it is developing technologies that will enable computing systems to extract and encode

> information from dynamic and stored data, observations, and experience, and to derive new knowledge, answer questions, reach conclusions, and propose explanations. Enabling computing systems with machine intelligence in this manner is now of critical importance because sensor, information, and communication systems continuously generate and deliver data at rates beyond which humans can assimilate, understand, and act [...] Recently, a more powerful approach has emerged, with rule-based, symbolic and human-oriented approaches combined with large-scale statistical approaches that make explicit use of massive distributed data and information bases. These data/information bases are curated (e.g. machine-filtered or human-selected) and raw (e.g. as originally obtained and perhaps of unknown provenance); structured (e.g. tabular or relational) and unstructured (e.g. text documents, multi-media files); static (e.g. historical, unchanging) and dynamic (e.g. real-time sensor data). This explosion in available data/information, combined with the ready availability of inexpensive mass storage and ubiquitous, inexpensive, computation-on-demand, provide the foundation for entirely new machine intelligence capabilities. The technologies developed in the Machine Intelligence project will result in revolutionary capabilities in military command and control, intelligence, decision-making, and situational awareness/ indications and warning for a complex, global environment where traditional (e.g. nation-states) and non-traditional (e.g. trans-national) actors and new classes of cyber–physical–human threats have become the status quo." [DAR 11]

Will AI, like other technologies before it, in turn have a profound impact on armed conflict, military confrontation, the way we think about war and the way we do it, i.e. the art of war? The decades that have passed since World War II, and, more particularly, since the 1960s and 1970s, have enabled the emergence of a new approach to the art of warfare that integrates the contributions of the calculator, the computer, IT and global computer networking. As early as the 19th Century, armies had already taken on board the consequences of global networking (telegraphy and then telephony in the 20th Century) and were at the origin of army reorganizations, the creation of new forces dedicated to mastering these tools, new strategies and tactics.

The advent of computers as early as World War II, and shortly afterwards, the creation of computer networks and the Internet, brought new tools to the military, offered them new possibilities and opened up new opportunities. They imposed a re-reading of the art of warfare. At the strategic, operational and tactical levels, all these new tools, both "calculators" and "communications instruments" with their new modalities, brought new analyses and practices, while remaining in line with a sometimes centuries-old continuity. The principles of information warfare, for example, have been modernized more in form than in substance.

AI must therefore be considered from the angle of its relationship to the present, to the past, to a heritage that is both near and distant in time: it is a branch of computer science, no doubt inheriting, at least in part, from what has been built around it and from the "cyber world" in the military field. Next, we will consider potential splits.

3.3.3. *Applications of AI in the military field*

Numerous open access documents discuss the concrete, envisaged or imagined applications of AI in the field of defense and its application in crises and conflicts. These approaches consider AI:

– as a weapon;

– in weapon systems;

– in the various functions of armies (command, control, logistics, maintenance, intelligence, including administration, etc.);

– for offense;

– for defense.

In 1983, as part of the Strategic Computing Initiative (a 10-year, $1 billion program), DARPA identified AI applications in various military fields. A matrix cross-referencing military applications and AI functionality was presented at a symposium in October 1983[16] (Table 3.3).

16 Proceedings of the symposium: https://www.cia.gov/library/readingroom/docs/CIA-RDP86M00886R000500040010-5.pdf.

	Autonomous vehicle	Battle Management and assessment	Pilot's assistant	Homing terminal	Automated design and analysis	War gaming
Vision	x	X		x		
Speech		X	x			
Natural language	x	X			x	x
Merger information	x	X	x			
Planning and reasoning	x	X	x		x	x
Signal interpretation	x	X	x			
Navigation	x			x		
Simulation/modeling	x	X			x	x
Graphics/ display		X	x		x	x
DB/IM/KB	x	X	x	x	x	x
Distant Communication	x	X	x			x
System control	x	X	x	x		

Table 3.3. *Reproduction of the matrix published in [ECK 84]*

AI now finds applications in a wide range of domains:

– robotics (smart robots);

– design of an underwater robotic system for underwater mine detection and anti-submarine operations as part of a DARPA project;

– deployment of troops;

– intelligent drones, to monitor borders and identify potential threats;

– target recognition: DARPA's TRACE (Target Recognition and Adaption in Contested Environments) project uses ML to automatically locate and identify targets using SAR (Synthetic-Aperture Radar) images;

– maintenance, logistics. IBM has deployed its Watson application in the U.S. Army to anticipate technical failures in Stryker combat vehicles;

– C2, C3I: decision support (all levels, strategic, tactical, operational); force readiness assessment, capacities; planning, operations; data integration, multiple sources;

– identification of terrorist propaganda on the Internet;

– image recognition applications, image analysis: surveillance, autonomous vehicles, mine-clearing;

– expert systems;

– electronic countermeasures;

– conflict simulation, from combat;

– simulation, tactical and strategic thinking tools;

– decision support, support for C2;

– intrusion detection (cyber);

– Offensive Counter Air (OCA), mission planning process;

– tactical planning;

– operational planning;

– enhancement of surveillance, planning, logistical support, decision-making and combat;

– AI-directed guns (Kalashnikov);

– army health systems, battlefield medical assistance, care for the military (Military Medical Record Data Analysis System);

– combat simulation and training systems (recent developments by the U.S. Navy with companies such as SAIC, AECOM, Orbital ATK; and by the U.S. Army with companies such as SAIC, CACI, Torch Technologies, Millennium Engineering);

– processing of large volumes of data of a heterogeneous nature. AI can process these data more efficiently;

– self-driving military vehicles;

– UAV control systems.

According to MarketsandMarkets[17], military AI developments in the coming years should focus on the following areas:

– warfare platforms (especially weapon-embedded AI, combat systems less dependent on human action, self-designed weapons capable of carrying out attacks in a collaborative manner);

– cybersecurity: the protection of armies' computerized systems, whether networked or not, is exposed to cyberattacks, which the deployment of AIs could eventually counter more effectively than human action or current cybersecurity systems;

– logistics and transport: the use of AI could help to optimize these functions, reduce costs, shorten lead times, detect anomalies and manage by greater anticipation;

– target recognition;

– battlefield healthcare (development of Robotic Surgical Systems (RSS) and Robotic Ground Platforms (RGP));

– combat simulation and training tools;

– threat monitoring and situation awareness are based on SRI operations that can mobilize AI-equipped equipment (surveillance drones);

– information processing;

– these topics or areas also constitute the market segments of the AI military market, which is expected to be dominated in the coming years by the "data processing" segment, according to MarketsandMarkets. Growth is expected to be strongest in China.

3.3.4. *Expert systems in military affairs*

The expert system is the crossover of AI methods for processing data and human expertise. It is therefore a place of encounter or cooperation between humans and machines. One of its main qualities lies in its low profile, giving the impression that it can master complex situations with the help of the machine: "Expert systems create islands of certainty in a society characterized by its rapid change, insecurities and unpredictability." [LID 96]

17 https://blog.marketresearch.com/8-key-military-applications-for-artificial-intelligence-in-2018.

Title or object	Year
A Combat Battle Damage Assessor Expert System [NEL 84]	1984
Expert Systems: Working Systems and the Research Literature [BUC 85]	1985
Expert Systems: Techniques, Tools, and Applications [KRA 88][18]	1985
Artificial Intelligence and Expert Systems for Government Executives [STE 86]	1986
Artificial Intelligence: Expert Systems for Corps Tactical Planning and Other Applications [BAC 87]	1987
SWAN: An Expert System with Natural Language Interface for Tactical Air Capability Assessment [SIM 88]	1988
An Expert System for Automating Nuclear Strike Aircraft Replacement, Aircraft Beddown, and Logistics Movement for the Theater Warfare Exercise [HAR 89]	1989
The Evolution of Artificial Intelligence and Expert Computer Systems in the Army [HAN 92]	1992
Expert Systems for Sea Mine Warfare [LU 94]	1994
An Expert System Opponent for Wargaming [CUT 95]	1995
An Example of How Expert Systems Could Assist Operations Planning [TAY 96]	1996
An Analysis of Using Expert Systems and Intelligent Agents for the Virtual Library Project at the Naval Surface Warfare Center-Carderock Division [LIE 00]	2000
Probe into Principle of Expert System in Psychological Warfare (ESPW) [LI 11]	2011

Table 3.4. *Some works and themes related to the use of expert systems in the military field (classified by year)*

The applications of expert systems have been very numerous, seeming to find their place in practically every field of activity:

– in banking and finance: “First USA Bank (Philadelphia, Penn.), a provider of Visa and Mastercard services, has installed a neural network-based credit card fraud detection system. The Falcon system, developed by HNC Inc. (San Diego, Calif.), employs neural network technology to learn and identify unusual transaction patterns that are indicative of fraudulent charges” [LIU 93];

– in the energy sector: “Talarian Corp. (Mountain View, Calif.), vendor of the RT works real-time expert system development tool, has signed a site

18 The book deals in particular with the applications of expert systems in military and security affairs: an expert system used in the identification of characteristics of terrorist groups; a prototype expert system in tactical air targeting; and finally an application in the field of tactical combat.

license agreement with Pacific Gas & Electric Co. (San Francisco, Calif.). PG&E intends to use RT works to operate and control its electric utilities system, as well as to improve its operating efficiency." [LIU 93]

The military field has not escaped this craze, and expert systems have found many applications.

The use of expert systems has been part of a relatively long period of military history (almost two decades), in the absence of more efficient technologies, on the one hand, and of the certainty that these systems contributed to the strengthening of military capabilities, in particular, on strategic aspects of the art of warfare such as anticipation, prediction, uncertainty, war fog, friction and force management, on the other hand. Expert systems contribute to military effectiveness through information literacy:

> "Integration of knowledge from sensors and the use of expert systems and artificial intelligence will allow sophisticated analysis of the enemy's intentions and the options to counter him […] Computer modelling of logistics problems and the use of expert systems will enable a better appreciation of the logistics requirements of the land force." [BAK 95]

3.3.5. *Autonomous weapons*

If there is a central concept today in the militarization of AI and its use in warfare, it is that of autonomy. Two notions are associated, that of the "weapon" and that of "autonomy".

The concept of an "arm" refers both to "a means (such as a weapon) of offense or defense"[19], and to the subdivisions of the armed forces, "a combat branch (as of an army)"[20].

At the beginning of the 20th Century, "autonomous weapons" was used to designate the subdivisions of forces:

19 https://www.merriam-webster.com/dictionary/arm.

20 *Idem*.

> "The creation of Aeronautics as an autonomous body and the return to the body they originally came from, the Artillery regiments, for defense against aircraft, including searchlight units, has resulted in the distribution between two arms directorates of all the means constituting 'Defense against aircraft' within the framework of military aeronautics: in aeronautics, night fighter planes and protective balloons; in artillery, anti-aircraft guns, searchlights."[21]

> "In the 16th Century, artillerymen formed a corps [...]. However, as these artillerymen [...] invented increasingly handy and practical arms, tried them out themselves, then introduced them into the infantry, the importance of the infantry increased from day to day: they raised it to the level of the cavalry and finally led the nobility to no longer differentiate between these two branches. This result was achieved from the 16th century onwards, when artillery, while reserving for itself the handling of complicated and powerful machines, began to develop independently and established the principle of its organization as a third branch, destined to strike decisive blows on the battlefield and to become the *ultimate ratio*." [BLO 98]

> "Mounted artillery, its role, its strength. The *Broad Arrow* [...] points out that mounted artillery should not be viewed as a stand-alone branch, but simply as an extra-mobile portion of field artillery."[22]

On the other hand, a distinction is made between the autonomous weapon and the auxiliary or support arm or branch: "[...] he emphasizes that the air force is no longer an auxiliary arm but an autonomous, powerful, high intervention arm."[23]

21 Projet de règlement provisoire de manœuvre de l'aéronautique. Part I, April 30, 1925, Imprimerie Nationale (Paris), available at: https://gallica.bnf.fr/ark:/12148/bpt6k9612769k.

22 *Revue du Cercle militaire*: bulletin des réunions d'officiers des armées de terre et de mer, Paris, Cercle national des armées (France), 1888, available at: https://gallica.bnf.fr/. ark:/12148/bpt6k57567663.

23 *L'Écho d'Oran*: Journal d'annonces légales, judiciaires, administratives et commerciales de la province d'Oran, April 2, 1935, available at: https://gallica.bnf.fr/ark:/12148/bpt6k6734197m.

> "In the 17th Century [...] two branches then take part in the confrontation: the infantry [...]; the musketeers [...]. The cannons are not yet assembled as an autonomous weapon and are used, on the front lines of the troops, as a weapon to accompany the 'infantry'."[24]

Today, we speak of autonomous weapons to refer not to the subdivisions of forces, but to the armament itself (missiles, intelligent robots, vehicles, platforms, drones) where autonomy is defined as the capacity of the machine to act, react and decide without human intervention. The machine is not remotely controlled, it is not supervised. The autonomous weapon is the one that in combat selects targets, decides and once the decision to fire is made, fires. Other objects than weapons, in the strict sense, can be equipped with autonomy, such as vehicles, for example.

The concept of autonomous weapons has been dealt with in military affairs for several decades.

We have identified several publications that deal with this autonomy in military affairs.

Title	Date	Definition of autonomous weapon
DTIC ADA093954: Proceedings of the Open Sessions of the Workshop on Imaging Trackers and Autonomous Acquisition Applications for Missile Guidance[25]	1979	
DTIC ADA113908: Guidance and Control Technology for Highly Integrated Systems[26]	1982	On Autonomous Integrated Weapon Systems

24 Recueil de l'Académie de Montauban: sciences, belles-lettres, arts, encouragement au bien, Académie de Montauban, 1977, available at: https://gallica.bnf.fr/ark:/12148/bpt6k6532179k.

25 Proceedings of the Open Sessions of the Workshop on Imaging Trackers and Autonomous Acquisition Applications for Missile Guidance, GACIAC-PR-80-01, November 1979, available at: https://archive.org/details/DTIC_ADA093954.

26 Guidance and Control Technology for Highly Integrated Systems, AGARD Conference Proceedings, North Atlantic Treaty Organization, February 1982, available at: https://archive.org/details/DTIC_ADA113908.

Title	Date	Definition of autonomous weapon
DTIC ADA140369: A Preliminary Investigation on the Application of Robotics to Missile Fire Control [MCI 83]	1983	Where the application of AI and robotics to solve fire control problems is being considered; the objective is to evolve surveillance and target designation systems into autonomous fire control systems.
DTIC ADA147272: Rand's Experience in Applying Artificial Intelligence Techniques to Strategic-Level Military–Political War Gaming [DAV 84]	1984	This paper discusses the experiences of RAND, which uses AI (and, in particular, expert systems) in war gaming, and where autonomous systems are responsible for political and military decision-making.
DTIC ADA436516: Artificial Intelligence and Expert Systems [STE 86]	1986	
DTIC ADA276575: Standard Target Materials for Autonomous Precision Strike Weapons [WAL 93]	1993	Laser-guided bombs, precision strike bombs. Interest in these weapons was revived with Operation Desert Storm. These autonomous weapons need the geographic co-ordinates of their extremely precise targets. Insufficiently precise data are defined as useless.
DTIC ADA282715: Autonomous Weapon Guidance[27]	1994	Bombs or missiles with predetermined targets. These are bombs that are tracked to their target by systems such as GPS, laser radar, etc.

Table 3.5. *Selected publications dealing with stand-alone weapons*

3.3.6. *Robotics and AI*

Today, a shortcut is commonly made between robots and AI, as they seem inseparable. The modern robot seems to be of interest only in its coupling with AI. But "robot" and "AI" are two quite distinct objects, domains or categories, whose meeting produces "smart robots".

Moreover, robots existed long before AI, and their use in warfare has already impacted tactics and strategies. In the art of warfare, the robot precedes AI.

Unmanned weapons appeared as early as World War I. During World War II, the Germans used the robot Goliath, a mini remote-controlled tank filled with explosives with a range of 3 kilometers: "The secret of the Robot planes which have been attacking England may soon be known. Advancing

27 https://archive.org/download/DTIC_ADA282715/DTIC_ADA282715.pdf.

on Cherbourg Peninsula, United States troops have captured intact two robot plane launching sites, and experts are studying their secrets."[28]

During World War II, the Allies used drones, radio guided aircraft, to bomb German targets in Europe (operations Aphrodite and Anvil in particular). There was, however, no real autonomy in these "robots". Pilots were needed to take off from these flying fortresses filled with explosives, who, after having put the planes on the right route and handed over to operators on the ground, had to eject themselves from the aircraft.

World War II was a testing ground for these new kinds of weapons: "It is reported that the Germans are using robot machine-guns on the Western front. These are operated electrically by remote control."[29]

Combatants made extensive use of these new remote-controlled weapons known as robots, robot bombs and robot planes. Thus, these robots became very early killer robots. News of the war related daily attacks on London and the major European cities by German missiles. The number of casualties was incalculable. This mechanization of war was no longer only deadly for armies, it was also deadly for the civilian populations who were the first targets of these "robots". Thus, "robot", which, in the imagination or collective representations, was an assistant to humans, became simply a weapon, whose lethal power was impressive. The robot of that time was essentially a remote-controlled bomb.

The 1930s saw many experiments in robotics, including production of the "pilot robot" airplane. The press of the time reveled in these experiments and demonstrations, and we often learn, with photographs and a few words of explanation, that ingenious inventors developed humanoid robots capable of virtually all human activities:

> "The most perfect robot in the world is claimed by a young English scientist, who, after 14 years of labor and at a cost of nearly $18,000, has at last completed a mechanical man which

28 "Robot planes' secrets, London", *Narandera Argus and Riverina Advertiser,* June 23, 1944, page 2, available at: https://trove.nla.gov.au/newspaper/article/101547657?searchTerm=Robot+planes (article identified using the https://elephind.com/ engine).

29 "Robot Machine-Guns", *The Argus*, Melbourne, September 12, 1939, available at: https://trove.nla.gov.au/newspaper/article/11242794?searchTerm=Robot+Machine-Guns (article identified via https://elephind.com/).

> is almost human. This amazing robot can talk, sing, whistle, laugh or carry on a conversation (for a half hour at a time), tell the time and the day, fire a revolver and read the small print of a newspaper."[30]

Just a few years before World War II, articles were multiplying in the daily press about robots in war: "Robots making war by machinery" (1927)[31], "The robot goes to war" (1934)[32], "Robots and war" (1935)[33], "Robot war may mean war's end" (1935)[34] and "Inventing weapons for robot wars" (1936)[35] are only a few examples of this dense literature.

These (sometimes fictional) accounts, analyses, reflections or testimonies of the technological advances of the time imagine the war of the future, which some people sense is very close. They imagine a war in which men are no longer the combatants. War then becomes only a matter for generals in command of their armies of robots. Victory goes to the robot army:

> "[…] the fleet commander […] snaps over a switch on the huge control-board before him. He is directing, by wireless, the fleet of crewless 'ghost ships', tiny boats laden with explosives, with which he is making his attack on the enemy fleet. […] The commander is quickly on the scene of battle – a battle which he has fought without men or guns, and far out of sight. The enemy fleet lies shattered […] The robot fleet is the victor."[36]

30 *Cambridge Sentinel*, Volume XXVIII, No. 42, October 15, 1932, available at: https://cambridge.dlconsulting.com/cgi-bin/cambridge?a=d=Sentinel19321015-01= en-20--1--txt-txIN-robot------- .

31 "Robots making war by machinery", *The Sun*, August 20, 1927, archival article accessed from https://trove.nla.gov.au.

32 McKay H.C., "The robot goes to war", *The Daily Telegraph*, Sydney, March 24, 1934, archival article accessed from https://trove.nla.gov.au.

33 "Robots and war", *To the editor of the Herald*, *The Sydney Morning Herald*, November 5, 1935, archival article accessed from https://trove.nla.gov.au.

34 "Robot war may mean war's https://trove.nla.gov.au", *The Daily Telegraph*, Sydney, June 27, 1935, archival article accessed from https://trove.nla.gov.au.

35 "Inventing weapons for robot wars", *The Northern Miner*, May 19, 1936, archive article accessed from https://trove.nla.gov.au.

36 "The robot goes to war", *The Daily Telegraph*, Sydney, March 24, 1934, archive article accessed from https://trove.nla.gov.au.

The research on and the craze for the robot of the 1930s were expressed in a different way during World War II. But men put a lot of hope and belief in it. We see, in the robot, one of the keys to the outcome of the war: "Germans captured in Italy believe they are on the way to victory through the robot plane."[37]

Or, on the contrary, the analyses are tempered: "The robot planes may do some terribly devastating work, but those who are most capable of judging say they cannot affect the ultimate issue."[38]

What would later be renamed short, medium and long range "drones" or "missiles" were then termed robots. These new capacities resulting from the world war would be important variables of new international relations:

> "'Robot bombs have sounded the death-knell to American isolationism, in the opinion of Government leaders', states the Washington correspondent of the Chicago Sun. They foresee a peace which must ban the use of such a weapon. The robot itself, and the likelihood that its range can be increased to span oceans, is regarded as the greatest argument for international agreements which has come out of the war."[39]

But we must not confuse unmanned weapons with autonomous weapons, which are not remotely controlled. "Robots" in the form of drones returned to warfare practice in the 1990s with their use in the Balkan War in 1995 (Predator drones), made possible with the advent of GPS. In the 2000s, the era of the combat drone took a new turn, with the use of attack drones in Afghanistan in October 2001 (killing Mullah Omar, leader of the Taliban) and then again in 2002. Other robots, notably mine-clearing robots, were used in the war in Iraq by the Americans (2004, American mine-clearing robot Sword). This robot would then be equipped with a weapon. The new American combat doctrine aimed to use as many automated combat systems as possible, which meant robotic and intelligent systems. In 2008, 12,000 robots were being deployed by the Americans[40]. In the same year, a sentinel

37 *The Newcastle Sun*, July 4, 1944, available at: https://trove.nla.gov.au/newspaper/.item/167588109?searchTerm=%22robots%20war%22&searchLimits=.

38 *Riverine Herald*, Echuca, June 20, 1944, newspaper available at https://trove.nla.gov.au.

39 "Ban on Robots after War", *The Argus*, Melbourne, July 14, 1944.

40 https://www.youtube.com/watch?v=x3fj92PIiWQ.

robot, a killer robot, was installed at the border between the two Koreas, the SGR-A, developed by Samsung, nicknamed "the indefatigable" because it operated 24 hours a day. The robot was armed with a machine gun and a grenade launcher. It was not autonomous. It located individuals and a remote human operator decided whether or not to engage the target.

The global market for military land robots could reach $10 billion by 2021[41].

The intelligent robot is vulnerable on several levels, just like any other tool/device or computerized object:

– on its material, physical dimension: it can be the object of attacks, of physical destruction; it is also primarily dependent on its electric energy sources;

– on its computational dimension: the input data can be altered, the algorithm can be fooled (e.g. telephones offering facial recognition but whose application is fooled by 3D printing); the algorithm is not error-free or free from "bugs"; the algorithm is a mathematical formulation, possibly flawed, with possible errors; the code is the translation of the algorithm, also subject to flaws, to errors and to software attacks; the output may be marred by margins of error; the learning phase is not invulnerable, nor is the learning phase from "prediction, interpretation";

– on the purely informational field: are the data interpreted, collected and then produced infallible, invulnerable, beyond dispute? AIs make it possible to produce information, images, sound, video, etc. and on these specific aspects, AI can contribute to luring, deception, manipulation, etc. But it can also be the victim of luring and manipulation. Is it able to protect itself?

3.4. AI and cyber conflict

It should be recalled that cyberwarfare can be defined as the exploitation, in times of armed conflict or in the context of armed conflict, whatever its phase (preparation, conduct, immediately afterwards), of cyberspace for offensive and defensive purposes (protection of military systems against enemy attacks, protection of non-military infrastructures, populations,

41 *Idem.*

anticipation and response to attacks, etc.). "Cyberwar" is the expression of armed conflict in the cyberspace dimension.

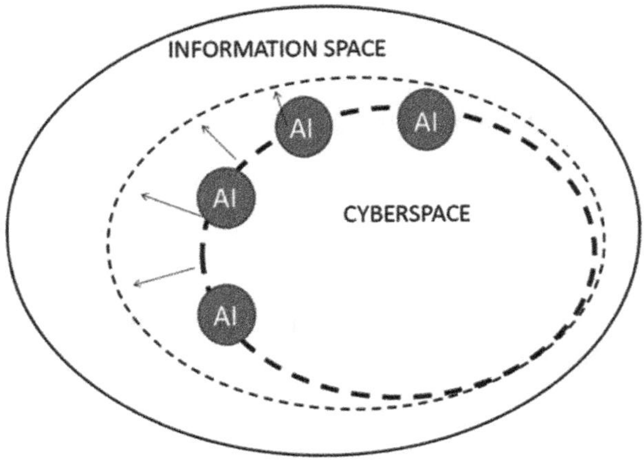

Figure 3.7. *AI is a component of cyberspace, which is itself a subset of information space. AI contributes to the expansion of cyberspace*

AI, for its part, as we pointed out in the introduction to the book, is a new aspect of cyberspace, one of its evolutions, which contributes to its expansion. AI is an integral part of cyberspace, which is a subset of information space. Even though cyberspace, due to its constant evolution, occupies an increasingly large place in the information space, it is still only a subset.

How does AI figure in military doctrines or strategies related to cyberspace? We have already pointed out the absence of any reference to AI in a key Department of Defense cyberspace strategy document, Joint Publication JP 3-12 "Cyberspace Operations". In its 2018 version [JCS 18], AI, as well as expert systems, Machine Learning, Deep Learning, robotics and the notion of "autonomous systems" or "weapons" are never mentioned. The 2013 version [JCS 13] did not mention them either. This silence surrounding AI in a document important for cyber defense can be interpreted as a desire to decouple issues that would be specifically cyber from another set that will be issues with AI. Unless, on the other hand, this implies a total integration of AI in the cyber domain, not requiring any mention of it. This approach of not mentioning AI in the official doctrine of cyber operations is naturally prolonged in military intelligence. The cyber operation guide, produced in November 2018 [CSL 18] for students at the U.S. Army War College, thus sets out the operational aspects of military action in cyberspace without ever mentioning AI.

The US Cyber Strategy for 2018 [DOD 18b], also published by the Department of Defense, does not use the concept of AI (or related concepts), either.

AI has recently (2019) been the subject of a formal doctrine for military AI in the United States [DOD 19]. In fact, this document also does not establish a strong, special relationship between AI and cyberspace. The latter is mentioned only four times in the document, notably to decide that:

– cybersecurity should help to secure AI. To this end, special efforts will be made to develop defensive cybersecurity tools;

– AI will be a security tool, improving the ability to predict and respond to cyber and physical threats.

In February 2019, the Director of the Joint Artificial Intelligence Center (JAIC), an organization whose creation was announced in the military strategic document on AI, spoke of his plans to work jointly with the Cyber Command (as well as with intelligence in the framework of the Maven project). The center's projects should deal with intelligence fusion tools, and the use of AI for C2 and cooperation with Cyber Command could cover incident detection and network mapping, for example. But here, again, the link with the domain "cyber" is not specifically highlighted. AI is, in any case not explicitly absorbed by the cyber domain, and Cyber Defense officials have not been charged with the responsibility of integrating AI into the defense forces.

If it is not in official published documents that we find a specific link between AI and cyberspace, then we must look for it in other places, such as statements or speeches by US defense officials.

3.4.1. *Malware, cybersecurity and AI*

Cyberattacks are now considered one of the most significant strategic challenges. Stealing data, manipulating it, modifying it, altering or prohibiting the course of data, changing the way data (and thus the systems that process them) are processed, etc., are all risks that can be taken into account in the fight against cyberattacks. The operational modalities are multiple and now well-proven. In this conflicting landscape, AI has made its appearance. Any immediate or potential change in both offensive and defensive capabilities in cyberspace is likely to have a strategic impact.

AI can be used in cyberattacks as well as in cybersecurity and defense. We thus find ourselves in the classic configuration of a duel, confrontation and balance of power. A balance is at stake, which each player tries to tip in his favor. Each duelist must then consider the benefit of AI: should it be used or not, why and how? The most important thing is not AI itself, but what it gives to each person. There may always be a temptation for experimentation or technical feats, but beyond that, it is important to think about its use more strategically before considering its implementation:

– Are attacks with AI controllable, both by their designers and by security measures?

– Is cybersecurity in any way incapable of countering attacks from AI?

– In a duel between two AIs (attack and defense), and in a duel where only one of the two is an AI, how are the forces distributed, what is the configuration of success (either for attack or for security)? Does AI guarantee superiority in all circumstances and in all environments?

– What non-AI strategies or tactics can be successfully opposed to AI? Is it correct to say that for the defender, it does not matter whether the attack is being or has been carried out by an AI, by a human or by both? [DAR 18]

– Are the means to be implemented to secure or attack by AI proportionate to the issues? For example, is it useful to deploy AI to attack systems that are insecure, poorly secured or very secure?

– Should AI be involved in all forms of attacks, in all phases of an attack?

– Does AI allow the invention of new forms of attacks or security?

– Are the means to be implemented within the scope of an actor?

AI would be synonymous with greater complexity of attacks, challenging security, as well as, conversely, making security more complex, making it more difficult for attackers to circumvent it. In both cases, complexity is then synonymous with quality and efficiency. But complexity, even if it increases, does not erase the duel and the search for an advantage on both sides. Moreover, complexity is intrinsic to the cyber domain. Cyberspace actors are used to it, and it has been omnipresent in cybersecurity and defense discourse for many years. It cannot, therefore, be considered as a new variable in these fields simply because of the integration of AI. On the other hand, AI could prove useful in navigating an environment that is too complex to be dealt with by conventional means, which could have become insufficient. On the cybersecurity side, AI is then perceived as a "solution", complexity as a "problem" or a threat-friendly context. On the attacker's side, AI is a solution to a problem that might be the resistance of those systems to be attacked. But we also know that the complexity of cyberspace does not really benefit security: there are many vulnerabilities, and the complexity of systems is one of the reasons. For cybersecurity, therefore, AI should not only help to detect attacks early enough, but also to lift a corner of the veil of complexity in order to better manage one's own environment.

AI would bring to security, as to attack, automation, understood as an autonomous action and not as a pre-programmed automatic process: this specificity could then be exploited in the detection of attacks and in reactions, as well as during attacks:

– on the offensive side, detecting targets and flaws, defining operating modes to achieve objectives without human intervention in the course of the process;

– on the security side, detecting attacks.

AI would give the malware used in random, opportunistic attacks a quality that, until now, has been the hallmark of humans: an understanding of the environment, of the context. The idea is to break with static software, i.e. software whose behavior corresponds to what has been precisely programmed, and to create applications capable of adaptation, of understanding the context in which they evolve and of making decisions based on the evolution of the context and their analysis. They would "think about" or analyze the situation in the same way as a human hacker would; they would model their behavior on that of the target's environment in order to go unnoticed, not to provide indicators of abnormal behavior to detection systems, or even to individuals observing, monitoring or even targeting operations. AI moves tactics and strategy away from the use of brute force, frontal attacks and visible, spectacular, predictable and visible violence. We would be in the realm of subtlety and the search for invisibility and discretion. But what would be the point of substituting AI, i.e. software, for the human intelligence capable of this contextual analysis? What is expected of this replacement? We will also consider human operators who use AI to help them improve their attacks.

Companies and state services are experimenting with the use of AI in offensive tools. The motive is, of course, officially to understand processes to better ensure protection. But actors with specifically offensive missions are even further ahead in their approaches, testing, development, analysis and even aggressive use of AI.

To analyze the behavior of AI-enhanced malware, IBM has developed DeepLocker, a form of malware that remains dormant until it finds its target. This type of malware will exploit data from multiple sources and from nature, such as facial recognition data, geolocation data, social media data, etc. The aim is to keep the malware hidden in the shadows, undetectable, for as long as the search for the target lasts, and to trigger it only once the target has been found. In this sense, we compare its *modus operandi* to that of the sniper: only trigger the shot after making sure that the target is the right one, that it is engaged and cannot escape the attack, but, above all, remain invisible until the last moment. DeepLocker is designed according to a model known as the Deep Neural Network (DNN), which makes it possible to mask the nature of the target (individual, organization), the identity of the target (who precisely) and the mode of attack (how the target will be attacked).

The U.S. Navy's Naval Information Warfare Systems Command opened a challenge in 2019 that awards 150,000 dollars for the development of AI capable of detecting cyberattacks.

Most of the capabilities provided by AI to carry out cyberattacks do not fundamentally represent a real break with the past and the present: the principles listed above are known and are extended in AI's built-in actions, such as masking its identity, acting anonymously, carrying out undetectable actions, targeting specific targets, carrying out complex attacks (in the sense that they combine several actions, several instruments, several attack paths, etc.), etc. AI's ability to carry out cyberattacks is also based on the fact that it can be used to carry out a number of different types of attacks, luring, deception, etc.

What they all have in common is that they all *take risks that are no more controllable today than they were before*. This makes them much like apprentice sorcerers playing with fire. How many malwares developed by agencies have ended up in the hands of the State or criminals, because they could not be kept under the control of their own designers? This is undoubtedly the real primary risk, more than the revolt of intelligent machines able to independently and voluntarily escape all control, and turn against their designers and their designs. The risk, as for more conventional malware and cyberattack tools, is data leakage and theft. It should be remembered that tools developed by the NSA were recovered a few years ago by hackers and, once modified, were reused to carry out new attacks that the United States itself and many other countries have paid for. Controlling an arsenal of any kind is a challenge. In many areas other than cyberspace, the question of protecting weapons and military equipment arises: weapons and equipment can be stolen, resold on the black market, recovered in theaters of operation or even stolen from R&D centers and companies, etc. In many other domains of cyberspace, the question asked is how to protect weapons and military equipment. These stolen or lost weapons, recovered by adversaries, are then returned to their original owners, users or designers.

For the victim of the AI malware attack, what difference does it make? The effect on the victim can be the same, regardless of whether the attack was carried out by an AI or by other means: data theft, computer encryption, disruption of company operations, financial losses, damage to image, information operations, etc. AI or not, the victim's situation hardly changes, except afterwards, at the stage of complaint, investigation, an attempt to reconstruct the crime and hope of prosecution of those responsible. In the post-attack phase, and especially after the attack is known (not all operations are intended to appear in the open; in the case of data theft, operations may, on the contrary, seek total invisibility and absence of any manifestation), the victim, the cybersecurity services, the police services, the justice system, the State or the actors in the counter-attack will encounter difficulties which are already known, but which will increase: the problem of allocating, locating and evaluating losses and damage suffered, the level of depth and gravity of the attack, the scale of the attack (which may have gone beyond the known perimeter of the victim alone) and the more conventional obstacles of the absence of borders in cyberspace for conducting operations versus the existence of strong national borders when it is necessary to act in the field of security, law, police action, justice or military action.

3.4.2. *AI and cyberweapons*

Conflictuality in cyberspace has taken shape by exploiting the multiple vulnerabilities inherent in this technological environment or created as a result of dependence on it. The search for and exploitation of vulnerabilities is the core of the cyber-offensive dimension (which, of course, must also develop strategies and tactics). On the defense and security side, the bulk of the work consists of anticipating attacks, which includes identifying vulnerabilities. However, active defense will go further by advocating counter-attacks, or even preventive or pre-emptive attacks. Defense then means knowing the targets' vulnerabilities. Whatever the position (attack, security, defense), vulnerability is central, and the "armory" in service of the offensive (by which we mean in a military as well as in a non-war context) is the essential tool of these operating methods.

The notion of "cyberweapons" calls for definitions that do not characterize it in a precise manner. It can be approached by analogy with the definition of a weapon.

A weapon is: "A tool that is used, or designed to be used, with the aim of threatening or causing physical, functional, or mental harm to structures, systems, or living things." [RID 12]

Let us recall the distinction that exists, however, between a weapon by nature and a weapon by use. The first is designed as an instrument of destruction, to cause damage, to exert force against an adversary. The second is any tool or instrument that is diverted from its primary use and can thus become an instrument of destruction: a sword is a weapon by nature; a piece of wood can become a weapon when it is used to strike an individual, it then becomes a weapon by use. In cyberspace, there are software tools that are designed to destroy or cause damage (e.g. Stuxnet), and other tools that were not created to contribute to offensive operations (e.g. the Ping command), but which can nevertheless be used by aggressors, armies or criminals, as part of their offensive actions.

A cyberweapon is, by analogy, with these definitions of a weapon: "A subset of weapons more generally: computer code that is used, or designed to be used, with the aim of threatening or causing physical, functional, or mental harm to structures, systems, or living beings." [RID 12]

> "Cyber means of warfare that are used, designed, or intended to be used to cause injury to, or death of, persons or damage to, or destruction of objects, that is, that result in the consequences required for qualification of a cyber operation as an attack." (*Commentary to Rule 103 of the Tallinn Manual 2.0*, cited in [WAL 18])

Following the distinction between weapons by nature and by destination, we will also consider two categories of cyberweapons.

The concept of cyberweapons raises many questions, including the line between what is and what is not a cyberweapon. Can means of infiltrating servers, stealing information or spying be considered weapons? Thomas Rid does not think so [RID 12], since espionage activities cannot be considered to be attacks with weapons, since the intention is not to destroy or cause physical damage to targets. This choice has legal implications, as it excludes espionage intrusions from the laws of armed conflict. This is reasonable, but questionable, as cyberspace has blurred these distinctions, and it is quite possible to consider that intrusion or interception attacks by foreign intelligence are carried out with weapons. For the use of weapons does not systematically mobilize the law of conflict, and may be, more simply, a criminal motive.

Weapons are also subject to control, prohibition and limitation measures.

The development of offensive and defensive instruments in cyberspace by the military is one facet of the militarization of cyberspace. The deployment and use of these cyberweapons is part of this militarization.

It is in this context of the use of cybernetic weapons by a variety of actors (state and non-state) and the associated strategic, political, legal and semantic questions that AI has recently emerged, seemingly opening the way to new capabilities.

3.4.3. *Offensive–defensive/security configurations*

Cyberspace is now commonly represented in the form of a space made up of three levels: its first stratum or layer is that of physical hardware; the second, that of applications (software, data); and the third, that of meaning, of information.

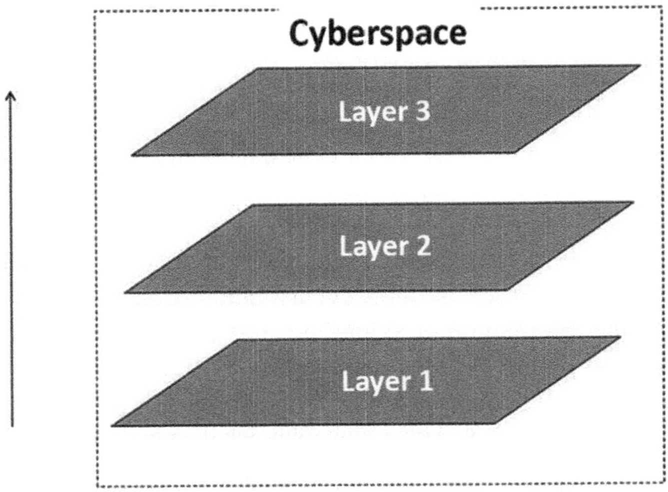

Figure 3.8. *A simple modeling of cyberspace in three layers*

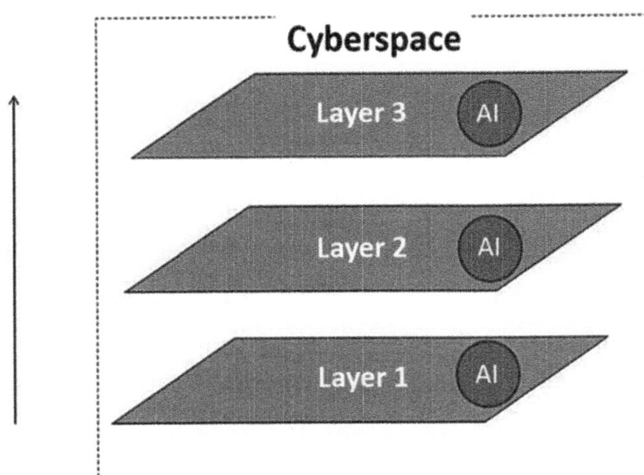

Figure 3.9. *AI is present in each of the three layers of cyberspace. It extends each of them and contributes to the expansion of cyberspace as a whole*

AI falls under each of these three levels. It is based on hardware, sometimes even specific hardware (s-chip, sensors, etc.); it is a form of software (programming languages, AI applications); and finally, it deals with meaning, it manipulates data, and, in its military applications, for example, its function is to provide human decision support, which falls expressly under this third layer. While AI brings new functionalities, allows data to be treated differently, involves particular software designs and is embedded in particular hardware (sensors, robots, etc.), it does not challenge this modeling, and, understood as an extension of cyberspace or as an element participating in its expansion movement, it must be applied to the same reading grid as cyberspace as a whole. We cannot imagine a fully fledged separate AI space or domain. We will not find reflections or propositions for the creation of a new AI space. We will thus remain within the conventional domains (land, air, sea, space, cyberspace) and will not consider a new spatialization built around AI.

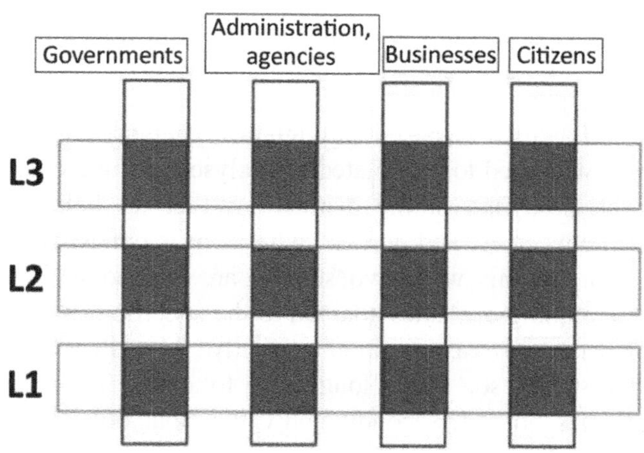

Figure 3.10. *Within each of the three layers of cyberspace, there are actors of different natures, with different functions, capacities, intentions and powers. For a color version of this figure, see www.iste.co.uk/ventre/artificial.zip*

At the level of each of the strata of cyberspace, different actors intervene, each with its role, its functions, its level of intervention, its intentions, its capacities and varying degrees of power. The weight of each of the actors is unequal, varying according to the strata in which they intervene. The State, for example, can influence the configuration of each layer, decide, authorize or prohibit developments and uses, regulate, supervise, militarize, etc. Not all States have the same weight on every layer. Companies are generally not actors in a position to intervene and influence each level, even if they are users. However, some companies are aiming to become global players, playing a strong role at the first level (by deploying international internet cables, for example), the second (by providing social media platforms, multinational software companies) and the third (by becoming information platforms). Citizens will generally be simple users at the first layer, can be users and developers at the second, and users/contributors at the third. The configurations, the balances, the weight of each of the intersections (in Figure 3.10) differ from one State to another, depending on regulation, law, governance, culture, organization of the State, the economy and society. The same will be true for AI, present on each of the three layers, and where each category of actor can play a role. Viewing the conflict in terms of the contributions of AI, or in which AI will be a target or a weapon or a vector of attack, will involve taking into account each of the intersections of the model in Figure 3.10.

However, AI, which is based throughout cyberspace, sharing common characteristics, will need to be isolated in analyses and in the development of tactics or strategies. One of the distinctions that we believe is useful is therefore that between AI and non-AI, what is or is not AI and what is or is not related to it. In this way, two subsets are created within cyberspace, facilitating the reading of tactics adapted to the specificities of each of them. There are at least two reasons for this: firstly, AI will not take up all the space in cyberspace, so, for a long time to come, it will coexist with cyberspace that is not AI; secondly, only the final objective sought must decide on the method and means (it should be remembered that AI is not necessarily always the best option, the most efficient, the one with the best return on investment, the one that proves to be the most rational, etc.). With the advent of AI, we have observed the emergence of specific actors, tools and practices that do not take ownership of all aspects of cyberspace.

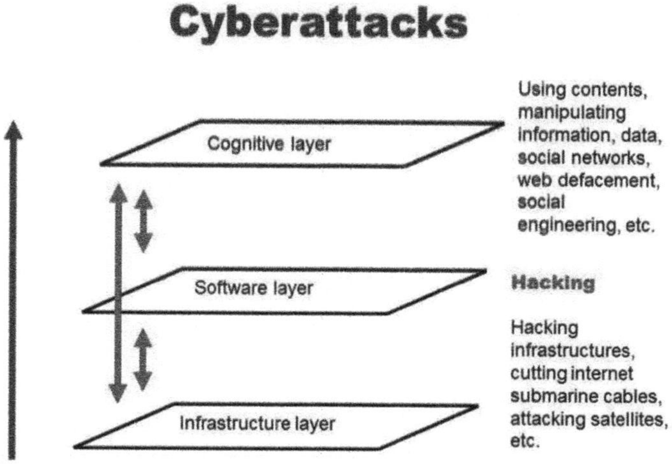

Figure 3.11. *In cyberspace, each of the layers has its own methods of hacking, attack and defense, and securing, to produce effects on one or more layers, inside and sometimes outside cyberspace. For a color version of this figure, see www.iste.co.uk/ventre/artificial.zip*

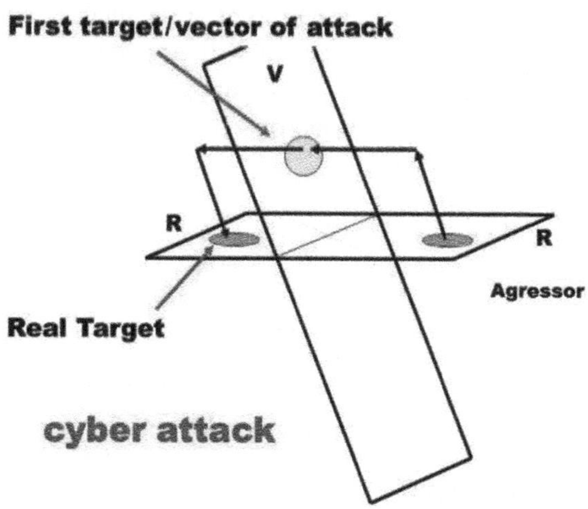

Figure 3.12. *If V represents cyberspace and R represents the non-cybernetic world, an attack is an action starting from R and hitting R. V is only a vector, an intermediary, a means of exerting violence, force, etc., and is not a means of achieving the desired result. For a color version of this figure, see www.iste.co.uk/ventre/artificial.zip*

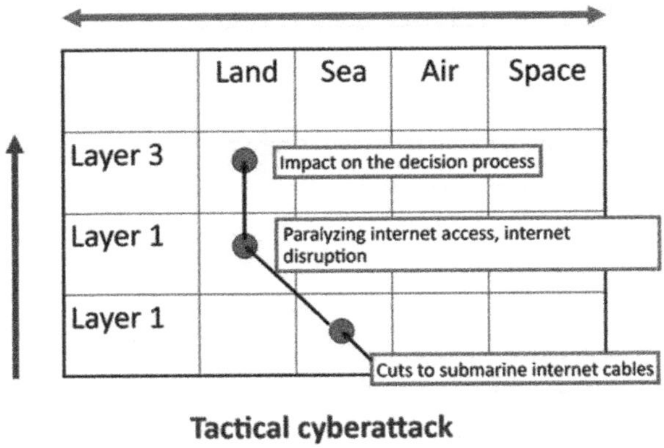

Figure 3.13. *An attack can exploit one or more of the layers of cyberspace to produce cascading effects. For a color version of this figure, see www.iste.co.uk/ventre/artificial.zip*

AI will not change the logic of the confrontation, which starts from a point (an aggressor, from the "real" world, which has a will and objectives) and aims at a target, also in the "real" world. A specific confrontation may appear, one that would oppose an AI-based resistance, security or defense, with an AI-based attack. In other cases, it will be a confrontation between an AI-based security/defense versus an attack without AI; or between an AI-free security/defense versus an attack based on AI; or a security/defense versus an attack, without any AI on either side.

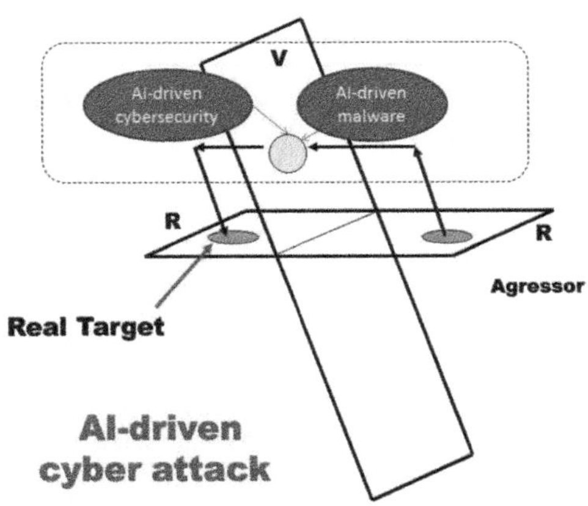

Figure 3.14. *Positioning AI in a cyberattack. AI malware will sometimes be opposed by an AI-based security/defense system. For a color version of this figure, see www.iste.co.uk/ventre/artificial.zip*

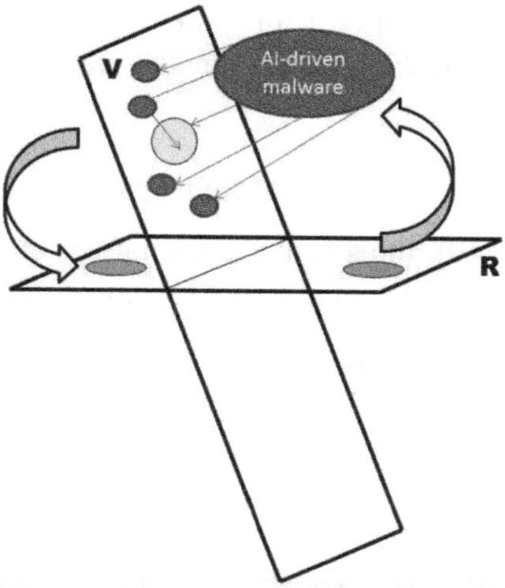

Figure 3.15. *Can an AI malware have a true panoptic or global view of its environment?. For a color version of this figure, see www.iste.co.uk/ventre/artificial.zip*

One of the specificities of AI-based malware is its ability to take into account its entire context or environment, and to self-adapt to this changing context. AI is therefore endowed with the ability to view an environment that surpasses that of humans, on the one hand, and, on the other hand, with a vision capacity that can be akin to a panoptic gaze. It is not a question of having a vision of cyberspace in its totality, but of the immediate context, and of transforming action by integrating modifications of the environment and of the target. An overview and real-time action capability gives AI malware the panoptic power over its field of action, as it can also remain invisible and undetectable while remaining in the closest proximity possible to their targets. A global view, from above, and action at close range.

In Figure 3.16, we represent some simple attack/defense configurations, integrating two basic situations: when the basic actor or system mobilizes AI (noted 1) or not (noted 0). Three actors and systems are considered here, simplifying an attack–security/defense scheme:

– the aggressor or attacker (noted marked Att);

– security or defense (noted Sec);

– the target, object, system or actor, (noted C) which, on the one hand, is attacked or can be attacked, and, on the other hand, is subject to security/defense.

We can read the paths shown in the figure as follows:

– Att (0)/Sec (0)/C (0): the attacker does not use an AI, the security does not use an AI, the target does not use an AI/is not an AI;

– Att (1)/Sec (1)/C (0): the attacker uses AI to carry out the attack, security uses AI to protect the "target", the target itself does not use AI/is not an AI.

This gives us eight paths.

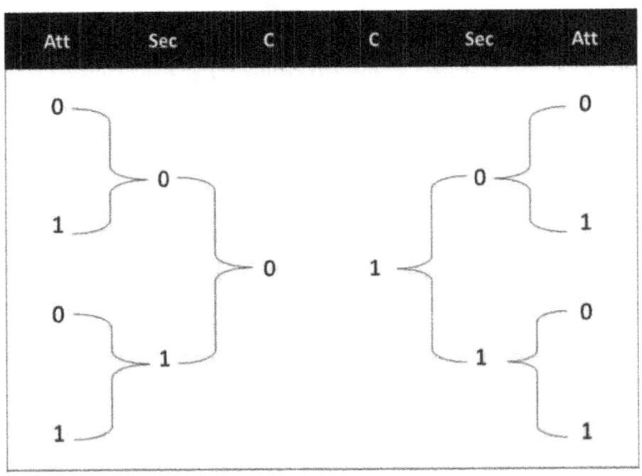

Figure 3.16. *Attack and secure/defend, with or without AI*

Some questions can therefore be raised on the basis of this reading grid:

– Is there a path by which the target cannot be attacked? On the other hand, is there a path where the target is totally vulnerable?

– What is the path that maximizes the security of C, the efficiency of Sec. and the effectiveness of Att.?

– When only one element mobilizes AI, who takes the advantage: the element using AI or, despite everything, the others?

– What happens when two or more AI forms are pitted against each other?

Responses will be determined by the quality of AIs, by their value or level, by their appropriate use and by the quality of their implementation. There will be some AIs of better quality than others, some higher or lower than others (a hierarchy of AIs), some adapted and some not to the objectives or missions. It can be assumed that AI is not an absolute advantage in all situations or configurations. It must also be considered that targets exposed to attacks mobilizing AI may eventually manage to devise alternative means of evasion, tactics or strategies that are native, involving AI or without AI. This is the principle of all asymmetric confrontations. Weak actors who cannot fight head-on look for avoidance tactics and strategies, bypassing tactics and other modes of operation that will take the aggressor from behind, on their weak points. Because the attacker, even with AI, will keep their weaknesses. The AI they use can be the first point of vulnerability itself (AI can be fooled or deceived).

This reading grid applies in the context of cybersecurity and cyber defense. The target can be a computer, server, computer, network, a connected weapon system, an individual or group of individuals, social media, etc.

Cybersecurity uses AI techniques to counter cyberattacks that may or may not exploit AI.

3.4.4. *Adversarial AI and adversarial Machine Learning*

DNI defines the notion of "Adversarial AI" as being: "A subset of AI focused on understanding how AI systems behave in the presence of a malicious adversary." [DNI 19]

Solutions using Machine Learning are developed to observe hacker behavior and adapt cybersecurity to changes in their tactics. Attack detection tools are becoming able to follow the changing practices of hackers, and are no longer limited to specific types of attacks. The new systems bring adaptability, flexibility and dynamics to the rigidity of the systems previously used. The objective is to detect attacks or attempted attacks much earlier in the attack cycle. However, hackers are also adapting their attack techniques, aware of the implementation of these new detection methods. The balance of power therefore remains intact. All the more so since ML systems are not infallible (when data is insufficient, so is learning). To counter intelligent security systems, hackers will find countermeasures, ways to deceive systems by posing as legitimate users, or will seek to understand how security systems work and act on the data used to learn the system. It is Therefore a question for the attacker is to hack the AI itself, to target its weak points, and deceiving an AI can sometimes be quite simple:

– AI used in the courts to decide the judgment is likely to be wrong. Indeed, having learned from the masses of data from previous judgments, it may have learned from errors in judgment. It is likely to produce them again. AI would copy the errors of the past [HAO 19];

– Microsoft's chatbot, Tay, launched on Twitter in 2016, quickly turned to hate speech. The AI had learned from content widely disseminated by trolls in an attempt to disrupt the functioning of forums. AI does not distinguish between what can be considered "good" or "bad", between what is legal and what is not., but this experience also reminds us that learning AIs can have their learning manipulated;

– systems of an autonomous vehicle that cannot detect the presence of another vehicle whose color is similar to that of the sky (an error causing a traffic accident);

– stickers on a road sign can be used to mislead a system of image classification (Deep Learning) [EYK 18];

– facial recognition systems have significantly high error rates and can be deceived [SHA 16];

– during the demonstrations in Hong Kong in August 2019, demonstrators used lasers and pointed them at the police force. The aim was to blind the cameras that feed data to facial recognition systems. When the camera is inoperable, the upstream recognition system is no longer useful. Intelligent systems are dependent on the data from their sensors: in order to foil AI, the sensors, the source of the data, must be attacked.

For the "attacker" of these AI-based systems, the method consists of attacking the data, which it captures and analyzes, and which is likely to make the AI fail to operate. Neural networks can be disrupted by signals or by data added to an image [SZE 13], or an audio signal, for example. These properties, which are vulnerabilities, have only been understood and identified relatively recently.

AI hacking is therefore becoming an issue both for attackers and for users of AI-based systems, whether in cybersecurity or in other areas such as robotics, autonomous vehicles, facial recognition systems, etc. The exploitation of AI vulnerabilities is of course a key issue in the military domain.

3.4.5. *AI and information warfare*

In *The Machine That Won the War* (Isaac Asimov, 1961), three leaders of the human race meet after their victory over the Denebians and discuss the role, decisive or otherwise, that the Multivac computer played in this success. The first admits that he changed the data entered into the computer because he felt he could not trust the quality of the reported data; the second admits that he changed the data produced by the computer because he felt that the computer was not in optimal mode; and the third admits that he did not trust the reports produced by the computer and flipped a coin to make decisions, relying on chance rather than intuition and the ability of the machine.

This story, although fictional, brings forth several questions that are central to the issues of information warfare and the art of warfare more generally:

– the place of the computer in warfare, in the exercise of C2, is entirely determined by the relationship between humans and machines;

– this human–machine relationship is defined in terms of trust:

- in the ability of individuals to ensure optimal machine operation,

- in the quality (accuracy, relevance) of the information/data produced prior to processing by a computer,

- in the technical capabilities of a computer,

- in the quality or relevance of the data produced by the machine;

– in the absence of trust in all these components that contribute to the decision-making process, the war leader in charge prefers to rely on his personal judgment, or even on chance.

These issues remain intact when considering the role of AI in command processes: the presence of AI does not guarantee the reliability of the data that supply it; AI itself may not be technically optimal and what it produces raises questions, even skepticism or concern (what does AI actually do with the data? How does it produce its results?); humans can choose not to rely on the data produced by AI when it is only a decision-support tool; and what guarantees that a decision taken on the basis of an analysis by AI is, in terms of consequences or effects produced, superior to a decision taken on the basis of human reasoning alone, or even on the basis of chance? AI depends on the data it is provided with, the data it processes.

AI aims to bring the computer up to the level of human intelligence, and to surpass the reasoning abilities of the latter. Information warfare, which plays on the level of knowledge, representations, reasoning, is *a priori* one of the major fields of AI application:

> "Compared to measuring material capability, assessing enemy intent is something close to a black art, requiring nearly clairvoyant insight into the personality and mental processes of the enemy commander. Artificial intelligence techniques of expert systems and pattern recognition may have application to this ancient problem and offer the prospect of developing into devastating new weapons of the information war. Using such techniques, it may be possible to build a machine analog of an opposing commander that could be used to test for reactions to various courses of action." [LEW 91]

One of the roles assigned to AI could therefore be to anticipate enemy decisions. AI could think like the enemy and put itself in their place:

> "Combining techniques of expert systems and pattern recognition, and perhaps other artificial intelligence techniques might, for example, lead to a machine simulation of an individual leader's expertise applied in a specific event environment to yield a 'most likely' prediction of that leader's response. Of course, if the leader in question is the enemy, capturing his expertise and identifying templates relevant to his pattern-driven behavior could also be massive tasks. The point is, such tasks are now merely very difficult, no longer impossible." [LEW 91]

A U.S. Army study report looks at the usefulness of AI in information warfare. Written by Colonel David C. Kirk [KIR 96], as part of a 1996 research project at the U.S. Army War College, the report, entitled

"Artificial Intelligence Applications to Information Warfare", argues that intelligent agents will in the future be key elements in information warfare to manage information flows, to build firewalls to protect networks and to make a major contribution to network security by assessing network vulnerabilities on an ongoing basis.

In this report, the author explores how the advances then made in AI could find applications in the conduct of information operations and in the security of information infrastructures. The author's reflections come at a key moment when the American defense forces are engaged in reflections on the impact of new technologies in military affairs, in the art of warfare and in the formulation of concepts related to information warfare. Let us summarize in a few lines the arguments put forward by the author:

– He begins by taking stock of the definitions of information warfare and, while stressing that this concept cannot be confined to the military field alone, it is the latter that guides his thinking. The definition adopted is that of the Department of Defense in 1994 [DSB 94]: information warfare refers to "actions taken to achieve information superiority in support of national military strategy by affecting adversary information and information systems while leveraging and protecting our own information and information systems"[42]. Information warfare refers to both defensive and offensive information operations. The concept of Command and Control Warfare (C2 Warfare) is the operational military translation of information warfare, which consists of attacking/defending the opposing information infrastructure/defending one's own/the opponent's information infrastructure in order to alter or defend the decision loop.

– The defensive dimension includes all the "actions to ensure the availability, confidentiality, and integrity of reliable information vital to national security." [WHI 95]

– During the first half of the 1990s, the information dimension had not yet become a strategic pillar in the political arena. The 1995 Presidential National Security Strategy only dealt with three pillars: diplomatic, political

42 For an analysis of the concepts and doctrinal corpus developed by the United States since the early 1990s dealing with new information technologies in military affairs, and in particular a comparative reading of the multiple definitions of the concept of "information warfare", we refer the reader to: Ventre D., *La guerre de l'information*, Hermes Lavoisier, Paris, 2007. Also available in English: Ventre D., *Information Warfare*, ISTE Ltd, London, and Wiley, New York, 2009 (1st edition) and 2016 (2nd edition).

and military, but did not yet give a fully fledged place to the information space.

– The Defense Forces were fully aware of what was at stake. Their information infrastructure (DII – Defense Information Infrastructure) was made up of an impressive number of computers, satellites and cables (which, it was said, would enable them to circumnavigate the Earth 400 times). 95% of military communications then used public networks. For this reason, networks were exposed to offensive operations and their protection was complicated. The complexity arose from the interdependence of all these systems, civilian and military, public and private. As early as the 1980s, these networks and systems were tested by the first computer attacks (in 1988, infection of 6,000 machines worldwide in less than two hours; in 1986 and 1987, 450 US defense computers were compromised by a German student).

– The mid-1990s were marked by a new wave of optimism around AI. Many people believed it was resolutely on the path to success.

– During this period (the 1990s), the intelligent system or expert system was one of the best-known forms of AI. Expert systems were defined by the author of the report as "computer systems that emulate the decision-making ability of an expert. Experts or knowledge engineers program a set of 'condition–action' rules. The expert system, when presented with a situation, searches for the appropriate condition and returns the associated response". It was in the field of medicine (diagnosis) that expert systems found their first applications, before they were used in the military field (for logistics, maintenance, planning, etc.). The limits of expert systems lay in the quality of their programming, and in the static side of the knowledge that was entered once and for all in the system. They were strictly limited to their expertise, unable to understand that they are going outside their comfort zone when problems arise that are outside their area of expertise. Expert systems can produce erroneous answers to the problems they are confronted with.

– "Agents" have partially overcome these weaknesses, as they allow knowledge to be shared between several systems. "An agent is anything that perceives its environment through sensors and acts upon that environment through effectors." "Smart" agents are those that interact with human users and especially with other agents: "they must be able to function without a human–machine interface"; "they must go beyond task orientation and seek to accomplish objectives without being told how to do the task." Intelligent agents are defined as software tools: "instead of user-initiated interaction via commands and/or direct manipulation, the user is engaged in a cooperative process in which human and computer agents both initiate communication, monitor events and perform tasks." The agent's performance depends on its ability to learn about the interests of the user, as well as those of the agent community. Learning is at the heart of these intelligent systems, the objective being to acquire a level of competence.

– How can these agents then be used in the information war?

- They can provide tools to help manage the mass of information. Even rudimentary applications (e.g. the SIFT software – Stanford Information Filtering Tool) make it possible to filter and search for specific information in the masses of data posted by Usenet users. Software applications offer the possibility of sending messages via Personalink without knowing the recipient's address, with the software taking care of searching for the information and delivering the message. These "agents" are not, however, strictly speaking "smart". But the prospect, for military applications, is then to make significant progress, to offer tools to help military commands to manage, to cope with the ever-growing mass of data.

- They can be deployed for infrastructure security. Progress needs to be made to develop smart firewalls; to monitor the activities of system users by learning about their habits, behaviors, working environment; to assess the level of security using agents specializing in tasks (assessing the security level of passwords, authentication, access authorizations, etc.).

At the time of writing this report, AI's capacity is still insufficient to develop all these projects. The results are encouraging, but there are still many limitations. The perspectives opened up by the author cannot yet be implemented.

3.4.5.1. *AI and memetic warfare*

A meme is information that spreads, has an impact and persists [FIN 11], a contagious thought, or the phenomenon of contagion of certain ideas and beliefs. The concept was formulated for the first time in 1976 by Richard Dawkins [DAW 76] in his book *The Selfish Gene*. The essential characteristic of memetic conflict lies in the reproduction, the propagation – like viral propagation – of false perception. Social networks have contributed to the growth of the phenomenon. However, the memetic conflict, described as a non-scientific object [BUR 12], is integrated into reflections on information warfare in modern conflict. In the *Military Intelligence Professional Bulletin* of April 2010, Lieutenant Brian J. Hancock [HAN 10] suggests using memetics to understand the causes of insurgent movements and explain their roots: "A meme is essentially an idea, but not every idea is a meme. In order for an idea to become a meme it must be passed on – or replicated to another individual" [HAN 10]; "a meme is information which propagates, has impact, and persists" [FIN 11].

From the latter definition, we will retain the three essential characteristics of the meme for conflict: it must spread, have an impact and be persistent. It should be added that memes are sometimes negative ideas ("harmful perceptions" [MAT 98]), but they are not limited to this type, because there are also positive, humorous, constructive memes, etc. A meme is essentially characterized by the fact that it spreads from individual to individual, by imitation. A meme is either an idea, or an image, a behavior, or information.

Memetic conflict consists of creating ideas and propagating them. The Internet and social media are vectors of this dissemination. AI can be one of the tools, for example, by allowing the creation of messages, content (the meme taking different forms, the possibilities of creation are multiple: words, symbols, icons, images, etc.), and the use of the Internet, to adapt them to contexts and environments, to personalize them, to adapt them to their target audiences, to modify their form and route, the intensity of dissemination, according to the evolution of the context, the effects produced, etc. The memetic war is a war of influence, disinformation and propaganda. Memes are suitable for social media, because to be effective, they must be short and concise (1–10 words) [FIN 11]. These reflections on the role of ideas and their manipulation have been financed by the American defense in the framework of projects supported by DARPA. Robert Finkelstein cites some of these projects, dealing with an epidemiology of ideas, military memetics, social media in strategic communication and narrative networks. These projects are being carried out over the period 2006–2011. The aim of the Social Media in Strategic Communications (SMISC) project is to produce automated tools and techniques for the detection and identification of adverse memes, and to automate the means to counter these types of attacks. Military interest in memes began after 2001 as part of the war of ideas against terrorism.

Memetic confrontation or conflict must oppose neutralizing tactics to viruses of ideology. These viruses, like malware, are now also spreading over networks in cyberspace. And while memes are as old as civilizations, modern means of communication have increased their speed of propagation and their range of action. To try to locate them in the mass of data circulating on the Web, social media have used AI to identify memes (Facebook uses an application called Rosetta to do this). Students at Stanford University used Machine Learning to generate memes [PEI 18] (Danklearning application) in 2018, capable of generating images from a single image.

The use of memetic attacks poses a threat to States that may be targeted by asymmetric actors:

> "Though we cannot know now the extent to which memetic techniques will prove to be effective additions to the practice of psychological warfare over the next 25 years, certainly, as Mr. Glabus demonstrates, we must be alert to the developing threat. To enemies who cannot match us on the conventional battlefields of land, sea, and aerospace, the battlefield of the psyche will be a tempting alternative." [MAT 98]

Regarded as a "viral" attack, it thus joins two other categories of viruses: bacteriological and computer viruses.

3.4.6. *Example 1: the war in Syria*

In modern conflicts, new technologies, new armaments and new tactics or strategies are being experimented with. AI therefore naturally found its place in the Syrian conflict (war against Islamic State). Its multiple aspects, which cannot be reduced to military confrontation per se, must be integrated into conflict observation. In particular, the effects of the fighting, the violence suffered by the civilian population, humanitarian considerations and the law must be taken into account. AI is gradually extending its field of action within the general framework of armed conflict, by investing in its multiple dimensions.

In the context of the Syrian conflict, AI has, for example, enabled the implementation of the Karim application, developed by the start-up X2AI, consisting of chatbots providing psychological support to Syrian refugees [SOL 16]. The application is based on natural language processing, a dialogue in Arabic that attempts to detect the emotional states of individuals in order to provide them with advice. The application has been publicized in refugee camps with the support of the association Field Organization Team. This application aims to support the mental health of refugees, many of whom have suffered emotional trauma.

A Machine Learning application was used to count casualties[43], and to list fake news about the war in Syria [ABU 19].

Facebook, for its part, is accused of having erased videos that could constitute evidence of war crimes. The algorithms, which are designed to automatically erase content identified as "terrorists", often contain videos showing the abuses being committed. Their analysis would make it possible to identify perpetrators, war crimes, discover the facts, know dates and so on. All this information, which is essential for the exercise of international justice, is thus automatically removed. The work of the International Criminal Court is now based on digital evidence[44] in the handling of cases involving armed conflicts. This work, which consists of collecting, archiving, analyzing and transcribing data from the Internet and social media, has been altered by the automatic content cleansing carried out by platforms such as Facebook and YouTube (YouTube removed around 90,000 videos per day worldwide in 2018). Of the 15 million extremist–terrorist contents deleted by Facebook between September 2017 and September 2018, 99.5% were deleted automatically by machines, and the remaining 0.5% were deleted by Internet users' requests[45].

AI has been integrated into military operations in Iraq and Syria [SAY 19]: in the Syrian context, the MAVEN project aimed in particular at identifying insurgents on the images collected by UAVs.

---

43 See the Human Rights Data Analysis Group website, https://hrdag.org/?s=Syriautton=Search and the video https://www.youtube.com/watch?v=B6xXeoo5QKA.

44 See, for example, the place of this digital evidence in the 2017 indictment file for crimes committed in Libya, available at: https://www.icc-cpi.int/CourtRecords/CR2017_05031.PDF.

45 The figures produced here are taken from [WAR 19].

3.4.7. *Example 2: events in Hong Kong in 2019*

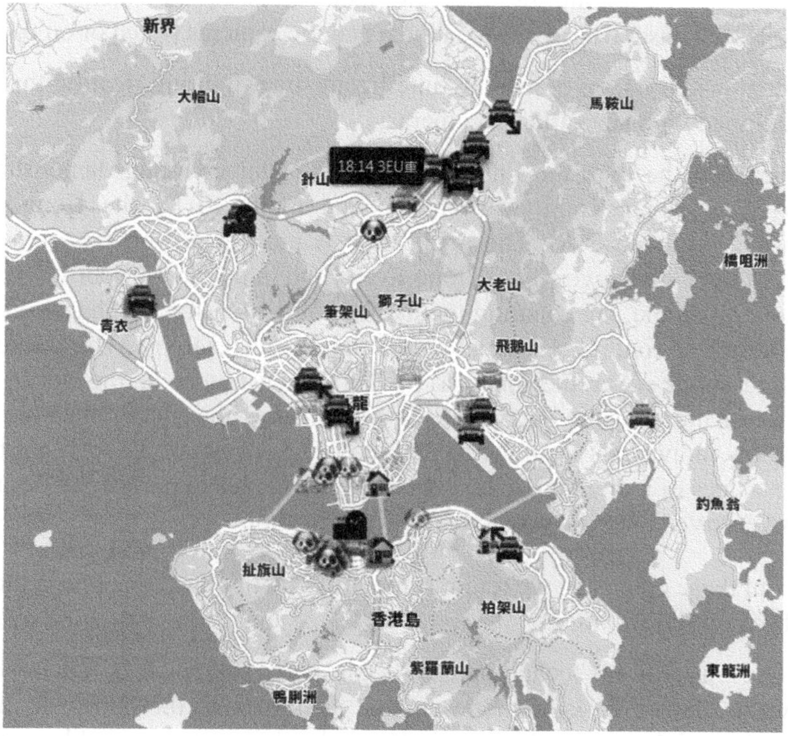

Figure 3.17. *Screenshot of the hkmap.live application (October 11, 2019). For a color version of this figure, see www.iste.co.uk/ventre/artificial.zip*

Images of the protests that took place in Hong Kong over the course of 2019 have traveled around the world. On one side, we see demonstrators occupying the public space *en masse*, on the other side, we see the forces of law and order trying to contain the crowds, to block the progress of the demonstrators. In the confrontation between demonstrators and police forces, each group uses their own strategy and adopts their own tactics. The technological arsenal in the field of information technology has been enriched by AI applications:

– the authorities, for their part, have deployed facial recognition systems, the aim being to identify demonstrators for law enforcement purposes (those considered rioters are liable to several years in prison);

– the demonstrators quickly adapted their behavior in response to this technological viewpoint, by masking their faces or by interposing their umbrellas and laser beams between them and the cameras (fixed or on drones);

– demonstrators are destroying intelligent street lighting, fearing that they are equipped with surveillance cameras;

– between the two, the Edge AI application for counting demonstrators, deployed in Hong Kong, was created and marketed by the Hong Kong company, C&R Wise AI. The application is used in universities across the country (e.g. City University of Hong Kong) and in public transport to manage the flow of people. By counting the number of demonstrators, the application can confirm or contradict official police figures, for example. The results can therefore serve either the interests of the demonstrators or the police. The founder of the company, Raymond Wong, ensures that the application does not follow individuals from the front, but only from the back, counting based on the shapes of individuals. There would therefore be no purpose or desire for intrusive surveillance, no possibility of identification.

From these elements, we retain some important points concerning the role of AI in conflict situations:

– like virtually all technologies, AI can be bypassed. It is not infallible, and weak players in asymmetric combat always have the ability to invent tactics from the weak to the strong. Here, the demonstrators held umbrellas to cameras;

– sensor-based AIs, and more generally AIs, are entirely dependent on input data. Here, by interposing umbrellas or masks, no data can be captured, AI is of no use, because it is not fed with data. AI is attackable at the level of the data and its sensors. And there is no need to oppose it to hacking practices or scientific and technological knowledge. It is simple changes in the behavior of the demonstrators that allow this resistance to AI;

– the physical, material extensions of AI (here, the cameras) are vulnerable to acts of destruction, sabotage and physical violence;

– crises or conflicts in smart cities take a different form from those in unconnected, conventional, non-modernized cities. The use of UAVs could be an alternative solution for security or surveillance and monitoring. But the images taken from UAVs do not allow the same images to be captured (in this case, faces). The future development of intelligent cities, the environments of connected objects, will on the other hand constitute for the inhabitants of these cities, spaces of surveillance from which it will be increasingly difficult for them to extract themselves;

– the use of AI complements, and does not replace, the now more conventional actions carried out in cyberspace, i.e. on the demonstrators' side, use of social networks, mobile telephony, livestream and a host of mobile applications, as well as tactics of influence; on the authorities' side, monitoring of social media, interception of communications, plans to partially shut down the Internet, etc.;

– the Chinese authorities have informed Apple that the company must remove from its Apple Store the hkmap.live application, which can be used to locate police forces and which they say may have been used by the demonstrators to escape from law enforcement on the one hand, and by criminals who attacked the police on the other. The American company, based in China, had no choice but to respond to the Chinese authorities' requests.

3.4.8. *Example 3: malicious AI attacks*

Scenario: fraudulent emails are spread by spam-phishing. But these emails are driven by AIs. Thanks to this "engine", each email is adapted to its target. It is contextualized, i.e. it is inserted into the flow of legitimate emails, automatically, from which nothing or almost nothing will distinguish it, which will make identification and detection by classic cybersecurity tools very difficult, if not impossible. The usefulness of AI in *malware* is here to make legitimate emails appear legitimate.

This form of attack by impersonating a legitimate person already exists in some ways, including identity theft. But with AI, the maneuver is on the one hand automated, on the other hand the attack is even more transparent. It is this ability to pretend to be something it is not, that malware exploits. Thanks to AI, the maneuver lures both humans and systems a little more. The practice will result in reducing the distinction between legitimate and illegitimate, with no or too little information differentiating the two in appearance. When that which is illegitimate has the appearance of the legitimate (chatbots pretending to be a human online, able to pass for a real interlocutor; deep fake, for the manipulation of image, sound and therefore perceptions), when that which is fake is as credible as the real one, how can we distinguish what should be put aside and what can be accepted? That which is legitimate can now be tainted by a suspicion of illegitimacy and danger.

These false, artificial contents, these decoys, are generated at the speed of the machine, diffused at the speed of light in optical fibers, and can therefore take any attempt to counter that which is strictly human. The speed of a human cannot rival that of the machine. However, in the manipulation of information, in disinformation, the one who produces a piece of information first has an advantage. Countering these discourses then imposes major efforts to contradict, demonstrate and reduce the effect produced by the first message.

For AI machines, generating credible content that looks like human-made content is not very complex. In social media, for example, messages written by humans are generally short (few characters, few sentences), poorly argued (no complex reasoning), sometimes conscious of spelling (natural language is simplified) and use symbols, abbreviations, "codes", which machines can easily learn.

3.4.9. *Example 4: swarming attacks*

When aggressors swarm, what forms of resistance, security and counter-attack oppose? Is swarming an invincible form?

The prospect of linking drones together, in very large numbers, to launch them against targets, is made possible by AI. Because with it, the communicating machines become a single unit, a single body, of which each part is part of the whole, and which can evolve, adapt, in its form and in the power it will exert against the target. The target can be quickly overwhelmed. Is it possible to oppose a swarm offensive with a swarm defensive? What protection can be offered to an isolated target against a swarming attack?

For the aggressor, several scenarios can be envisaged. They will configure their swarm according to whether the target will be isolated or a swarm.

The isolated target could be:

– with an AI:

  - unprotected,

- protected: with AI, or without AI,
- reactive,
- non-reactive;

– not equipped with an AI:

- unprotected,
- protected: with AI or without AI,
- reactive,
- non-reactive.

The swarm (let us consider that it imperatively embarks from AI) can be:

– unprotected;

– protected;

– reactive (counter-attack, resistance, resilience);

– non-reactive (passive, will not react against the aggressor and may be limited to resilience).

The swarm can constitute:

– of communicating individuals (a crowd, an army, etc.);

– of drones, robots, etc.;

– from malware;

– of sources of information in the case of an information war attack (cloud of websites, media, real or fake accounts on social media, etc.). AI facilitates the creation of virtual Internet users, automated content, adaptation of the content produced to the objectives and evolution of the context.

3.4.10. *Example 5: crossing universes with AI and without AI*

Just as the contemporary world can be divided into two groups, on the one hand the connected and cyberspace-linked spaces, and on the other hand the unconnected areas or those without links to cyberspace (the latter tending to shrink thanks to mobile communications, but still constituting a very large space on the planet), we can consider several universes in relation to AI:

– an environment devoid of any AI [AI0];

– a fully AI environment, all AI [AI1];

– a mixed environment, including some AI. The environment is not entirely dependent on AI [AI2]:

- this environment may result from a construction, for example, a system in which a few AI components are integrated,

- it can also be the result of an AI intrusion into [AI0] when, for example, an AI attacks a system that does not have an AI to protect it,

- in both cases, AI appears as a new element that will influence the environment with solely its presence, whether hostile, aggressive or not.

We will consider that:

– [AI1] is rare;

– [AI0] is probably dominant;

– [AI2] is gradually expanding.

Questions can then be raised about the balance of power and the balance of power between these circles:

– What will a confrontation between [AI0] and [AI1] look like?

– Is [AI1] > [AI0]? In the asymmetry between the two configurations, which can get the better of the other?

– Is [AI1] > [AI2]?

– In [AI2], does AI take over the entire environment? Does it dominate everything that is not AI?

Conclusion

The key concepts associated with artificial intelligence (AI), and thus constituting its domain or space, are those of speed, autonomy, invisibility, action, decision-making, learning, data, etc. With this particular AI space, we must associate its flaws and vulnerabilities, such as errors, biases, imperfections and exposure to attacks, which are the most important aspects of AI.

Many of these aspects are already present in the issues addressed by cybersecurity and cyber defense. But when we talk about cyberspace, we are primarily considering reticulation (networking of the world, individuals and societies), and we therefore treat security and defense issues in light of this essential characteristic. The network calls into question the way in which we control space, and therefore the way in which we think about national space, sovereignty, borders, the reduction of distances, power over this space and the militarization of this space. The network highlights the notions of flow, exchange and sharing. Cyberspace implies constant fluidity and permanent movement of data, and therefore is a dynamic associated with its own spatialization.

AI is placed at a different level; it calls for other concepts, other ideas, other representations and logics. Nonetheless, it is a full-fledged part of cyberspace, without which it is nothing: it feeds on data, is made up of applications, acts and circulates in cybernetic space and acts on cybernetic space, which it will modify, disrupt and reconfigure. Introducing new components into cyberspace with different behaviors (intelligent, autonomous, self-adaptive, capable of luring, imitating, blurring the line between true and false, real and artificial) is likely to reshuffle the cards.

AI is an integral part of cyberspace. It should therefore be treated as such, and be an integral part of cyberspace security and defense policies and doctrines.

National cybersecurity strategy	**Artificial intelligence**	**Machine learning**	**Robot**
National Cyber Security Strategy, Luxembourg, 2018[1]	4	0	0
National Cybersecurity Strategy, Spain, 2019[2]	2	0	2
National Cyber Strategy of the United States of America, 2018[3]	2	0	0
National Cyber Security Strategy 2018–2023, Jordan[4]	2	0	0
New Zealand's Cyber Security Strategy 2019[5]	2	0	0
Cybersecurity Strategy, Estonia, 2019–2022[6]	1	1	3
National Cyber Security Strategy for Norway, 2019[7]	1	0	1
National Cybersecurity Strategy, Republic of Korea, 2019[8]	1	0	0
National Cybersecurity Strategy, UAE, 2019[9]	1	0	0
Dubai Cyber Security Strategy, 2017[10]	1	0	0

1 https://hcpn.gouvernement.lu/dam-assets/fr/publications/brochure-livre/national-cybersecurity-strategy-3/national-cybersecurity-strategy-iii-en-.pdf.

2 https://www.dsn.gob.es/es/file/2989/download?token=EuVy2lNr.

3 https://www.whitehouse.gov/wp-content/uploads/2018/09/National-Cyber-Strategy.pdf.

4 http://moict.gov.jo/uploads/studies/National%20Cyber%20Security%20Strategy%202018-2023.pdf.

5 https://dpmc.govt.nz/sites/default/files/2019-07/Cyber%20Security%20Strategy.pdf.

6 https://www.mkm.ee/sites/default/files/kyberturvalisuse_strateegia_2022_eng.pdf.

7 https://www.regjeringen.no/contentassets/c57a0733652f47688294934ffd93fc53/national-cyber-security-strategy-for-norway.pdf.

8 https://www.krcert.or.kr/filedownload.do?attach_file_seq=2162ttach_file_id=EpF2162.pdf.

9 https://www.tra.gov.ae/userfiles/assets/Lw3seRUalMd.pdf.

10 https://csc.dubai.ae/res/wp-content/uploads/DCSS-EN.pdf.

National cybersecurity strategy	Artificial intelligence	Machine learning	Robot
Revue stratégique de cyberdéfense, France, 2018[11]	1	0	0
Singapore's Cybersecurity Strategy, 2016[12]	1	0	0
National Cyber Security Strategy, Turkey, 2016–2019[13]	0	0	0
National Cyber Security Strategy 2016–2021, Progress Report, UK[14]	0	0	0
National Cyber Security Action Plan, 2019–2024, Canada[15]	0	0	0
The Cybersecurity Policy for Critical Infrastructure Protection, Japan, 2019[16]	0	0	0
Cyber Security Strategy, Jersey, 2017[17]	0	0	0
Cybersecurity Strategy of the European Union, 2013[18]	0	0	0
National Cybersecurity Policy, Chile, 2017[19]	0	0	0
National Cyberspace Security Strategy, China, 2016[20]	0	0	0
Estrategia Nacional de Ciberseguridad *de Costa Rica*, 2017[21]	0	0	0

11 http://www.sgdsn.gouv.fr/uploads/2018/02/20180206-np-revue-cyber-public-v3.3-publication.pdf.

12 https://www.csa.gov.sg/~/media/csa/documents/publications/singaporecybersecuritystrategy.pdf.

13 https://www.uab.gov.tr/uploads/pages/siber-guvenlik/ulusalsibereng.pdf.

14 https://assets.publishing.service.gov.uk/government/uploads/system/uploads/attachment_data/file/805677/National_Cyber_Security_Strategy_Progress_Report.pdf.

15 https://www.publicsafety.gc.ca/cnt/rsrcs/pblctns/ntnl-cbr-scrt-strtg-2019/ntnl-cbr-scrt-strtg-2019-en.pdf.

16 https://www.nisc.go.jp/eng/pdf/cs_policy_cip_eng_v4.pdf.

17 https://www.gov.je/SiteCollectionDocuments/Government%20and%20administration/C%20Cyber%20Security%20Strategy%2020170215%20VP.pdf.

18 https://ec.europa.eu/newsroom/dae/document.cfm?doc_id=1667.

19 https://www.ciberseguridad.gob.cl/media/2017/04/NCSP-ENG.pdf.

20 https://chinacopyrightandmedia.wordpress.com/2016/12/27/national-cyberspace-security-strategy/.

21 https://micit.go.cr/images/imagenes_noticias/10-11-2017__Ciberseguridad/Estrategia-Nacional-de-Ciberseguridad-de-Costa-Rica-11-10-17.pdf.

National cybersecurity strategy	Artificial intelligence	Machine learning	Robot
Danish Cyber and Information Security Strategy, 2018[22]	0	0	0
National Cybersecurity Strategy 2017–2021, Egypt[23]	0	0	0
National Cyber Security Strategy, Greece, 2017[24]	0	0	0
Estrategia Nacional de Seguridad Cibernética (National Cyber Security Strategy), Guatemala City, 2018[25]	0	0	0
National Cyber Security Strategy for the State of Kuwait, 2017–2020[26]	0	0	0
National Strategy for Digital Security, Monaco, 2017[27]	0	0	0
National Cybersecurity Policy, Nepal, 2016[28]	0	0	0

Table C.1. *Searching for AI references in some recent national cybersecurity strategies*

However, we have observed that this is not the case. A separate path is being paved for AI. Militarily, cyber-commanders are not being given full responsibility for the militarization of AI. Specific structures are being created for AI. National cybersecurity strategies only cautiously integrate AI,

22 https://m.fm.dk/~/media/publikationer/imported/2018/strategi-for-cyber-og-informations sikkerhed/danish-cyber-and-information-security-strategy_weba.ashx.

23 http://www.mcit.gov.eg/Upcont/Documents/Publications_12122018000_EN_National_Cyber security_Strategy_2017_2021.pdf.

24 https://www.enisa.europa.eu/topics/national-cyber-security-strategies/ncss-map/GRNCSS_EN.pdf.

25 http://mingob.gob.gt/wp-content/uploads/2018/06/version-digital.pdf.

26 https://citra.gov.kw/sites/en/LegalReferences/English%20Cyber%20Security%20Strategy.pdf.

27 https://amsn.gouv.mc/var/amsn/storage/original/application/822de9d606448af4e900f566 abd3e00c.pdf.

28 https://nta.gov.np/wp-content/uploads/2018/05/Nepal-Cybersecurity-Policy-Draft.pdf.

which is usually presented as one element among others, into cyberspace. At the same time, AI is the subject of specific national strategies.

Few national cybersecurity strategy documents give any kind of line to AI. Jordan's strategy is one such document:

> "Cyber criminals are using AI bots to place more targeted phishing adverts and emails and analyzing large amounts of social media information to profile their targets. Online chat bots are also being seen more and more in use for customer service – positioning them as a system that people trust. Attackers will look to use this trust and build chatbots to try and obtain financial details from people."[29]

We will also mention Luxembourg's:

> "Threats inherent in the development of artificial intelligence. Generally speaking, artificial intelligence will emerge in the context of designing and executing malicious software. It is currently a research subject in the world of academia and will soon be mature enough to be implemented in real malware, which will have the ability to adapt dynamically to security measures put in place by teams defending the networks and infrastructure of organizations."[30]

In both cases, AI has been identified as one of the new forms of cyber threats. But nothing is envisaged in terms of the methods or means that will have to be deployed to counter them.

New Zealand's strategy[31], on the contrary, simply proposes a challenge to AI, and does not go beyond it: "A computerized system capable of simulating human decision-making and learning, including performing cognitive functions associated with the human mind including learning and language".

29 http://moict.gov.jo/uploads/studies/National%20Cyber%20Security%20Strategy%202018-2023.pdf.

30 https://hcpn.gouvernement.lu/dam-assets/fr/publications/brochure-livre/national-cybersecurity-strategy-3/national-cybersecurity-strategy-iii-en-.pdf.

31 https://dpmc.govt.nz/sites/default/files/2019-07/Cyber%20Security%20Strategy.pdf.

More generally, when these strategic documents mention AI, it is only to remind us that it belongs to a broader set of new technologies, such as cloud computing or the Internet of Things, which are destined to transform cyberspace and modern society.

The decoupling of AI-specific policies, on the one hand, and cybersecurity policies, on the other, is therefore a common approach around the world. Currently, AI is barely mentioned in cybersecurity policies. We have also noted the very weak presence of AI in US defense doctrine documents, even though the government and defense policy show a strong commitment to AI; US defense has invested heavily in these technologies and for a long time, and AI is one of the leading technologies in the modernization of armies. All this gives the impression of an uneven level of consideration for AI. Russia, in 2019, was still in the process of preparing its own national strategy for AI. The exercise has only recently been implemented all over the world. It will therefore still take a little time, a few months, probably a few years, before AI finds its full place in security and defense policies, as well as in military doctrines. To speed up these processes, it may take a more significant incident than others, fully attributable to the exploitation of AI capabilities.

A few points seem to grab our attention to conclude this work:

– AI is portrayed as a transformative, disruptive, revolutionary technology or set of technologies, but it has only recently been taken into account at political, strategic and doctrinal levels. The new wave of AI is only in its infancy;

– discourse is one of the main elements that triggers the different phases of AI. Thus, we have highlighted the role of negative or alarmist discourse, often emanating from researchers, whose arguments are heard by politicians and provoke, for example, the first winter of AI. However, the discourse of politicians also plays a decisive role: Ronald Reagan's 1982 speech marked a new period of huge investment in scientific and technological research;

– national mobilizations around major technological projects are also the expression of a reaction to the international environment. The security dilemma and the fear of loss of economic and industrial power are leading nation states into a technological development or arms race. This was, for example, the situation in 1982 when Japan announced its plan to create new-generation computers. The major powers reacted immediately and committed themselves to new R&D plans, both civil and military. The revivals that mark the history of AI, like other technologies, are the result of the position of the States in relation to the rest of the world. Today, competition between the United States, China, Russia and Europe is the driving force;

– the "militarization" of AI is written in its genesis. Technology is military or civil, or dual use. It is financed and its results are monitored by the military, in the United States or Russia, and in other countries, from its inception. It is in the same position as other technologies, such as the Internet and computers, in particular;

– with each generation, AI is reduced to one of its predominant technologies. There was an entire period marked by expert systems, the flagship technology of the 1980s and 1990s. Today, AI seems to be reduced to machine learning. But this simplification is probably useful;

– this reduction contributes, among other things, to feeding agreed AI discourses. Indeed, the stories we hear about policies for AI, AI safety or AI research are relatively simple. They tell us that technologies (here AI, but we find the same thinking when it comes to cyberspace or cybersecurity) are beneficial to society, must be developed and introduced, for the benefit of individuals (their quality of life and security). But the primary project that these technologies and their implementation must pursue is essentially economic in nature (forgetting, in fact, projects for networking citizens, planetary communication, access to knowledge or democratization of the world). However, obstacles are being put in the way of this project: enemies, adversaries and criminals are attacking society by attacking technologies or, with the help of technologies. We must then secure, protect and defend all of these technologies, because the survival of the system depends on them. We must also learn to master these technologies and turn them into weapons, so that we can respond to our enemies on equal terms;

– AI does not challenge the belief and the blind confidence that certain leaders, political and military, can have in technologies that are supposed to make them omnipotent. AI is a totem, a magical weapon that humans, who are eager to establish their power and might and to impose their violence, brandish and show to their populations, their troops and the rest of the world. AI is the extension of this belief or deep conviction that the universe, the world, the world of economics, politics, human government and war can be dominated by the mastery of information and, in the 21st Century, of data. The construction of this belief crosses centuries, but it took on a particular form in the 20th Century with the computer and networks. Today, with AI, data is, more than ever, the source of absolute power. These beliefs are never questioned. Indeed, official documents do not question the limits of this unbridled race for data, this dependence on data or the feasibility of projects that consist of controlling the world through data;

– AI, like cyberspace, appears to be indispensable to power in war. It is considered essential in building victory in wars. Before AI, there was the Internet, telecommunications, satellites, etc. With AI, there were expert systems; today, there is machine learning. However, the accumulation of these technologies, each time described as revolutionary, has never guaranteed victories or even absolute security. Opponents have always learned to adapt, bypass, do without the technologies or, on the contrary, appropriate them, building asymmetric or hybrid conflicts. It is unlikely that this new AI will guarantee military and political victories any more than yesterday's technologies did. The conditions of victory, as well as of economic and political power, are not reducible to a single technology. AI will not offer a faster victory either, even if it seems to compress time, which is, above all, that of data processing, calculation and algorithms. Yet, the temptation to remove humankind from the decision loop and to go faster seems to be great. Victory would only be possible at speed. However, if AI increases speed of action or decision-making, it still needs time to learn, and it also needs to be able to communicate, for the time being at least, with the rhythms of the human who decides its fate and activation. This obsession with time reduction and the exclusion of humankind, considered too slow or unable to process the huge masses of data produced by systems that they themselves have created and deployed, reflects a logic that is more productivist than productive: the machine would be more efficient, more effective and faster than the human. War would then be reduced to a question of managers, and human resources would be the last weak link in the business of war.

Appendices

Appendix 1

A Chronology of AI

In our proposed timeline of AI history (Table A1.1), phases of armed conflict and important political events are highlighted (shaded rows and columns).

Milestones in AI history	Year	Country
Start of World War I[1]	1914	
End of World War I	1918	
Play by Karel Čapek, *R.U.R.* (introduces the term "robot")	1921	Czechoslovakia
Beginning of the USSR (December 30, 1922)	1922	USSR
Metropolis movie, whose heroine is a female humanoid robot	1927	Germany
Design of Japan's first "robot", Gakutensoku	1929	Japan
Beginning of World War II	1939	
American scientists W. McCulloch and W. Pitts attempted to create an artificial algorithm of neurons	1943	United States
End of World War II	1945	

1 In this timeline, we have highlighted the dates of armed conflicts in gray. This does not mean that these conflicts were marked by AI advances and technologies. We indicate them here more as markers of different phases of history. The choice of conflicts is also debatable, as many other similar events have marked the course of world history during these periods.

Milestones in AI history	Year	Country
Beginning of the First Indochina War (December 19, 1946)	1946	Indochina
Norbert Wiener published his book *Cybernetics: Or Control and Communication in the Animal and the Machine* (the first public use of the term cybernetics)	1948	United States
Dominique Dubarle published in the newspaper *Le Monde* the article "*Vers la machine à gouverner*", devoted to cybernetics (after Wiener) (December 28, 1948)	1948	France
Edmund Berkeley published his book *Giant Brains: Or Machines That Think* (in which he compares early calculators to human brains)	1949	United States
Start of the Korean War (June 25, 1950)	1950	Korea
Turing's test. Article "Computing Machinery and Intelligence"[2] (published in the philosophy review *Mind*) (proposes the now famous test or imitation game)	1950	United States
Claude Shannon published the article "Programming a Computer for Playing Chess"[3]	1950	United States
Marvin Minsky built SNARC, the first neural network simulator	1951	United States
End of the Korean War	1953	Korea
Beginning of the Cuban Revolution	1953	Cuba
End of the First Indochina War	1954	Indochina
Beginning of the Vietnam War	1955	Vietnam
Logic Theorist, the name of what was considered the first AI program. Created by Herbert Simon, Allen Newell and John Shaw	1955	United States
The MANIAC computer played the first chess game program	1956	United States
Dartmouth College Conference organized by John McCarthy, Marvin Minsky and Claude Shannon. Introduction of the phrase "artificial intelligence" by John McCarthy	1956	United States

2 Turing A.M., "Computing Machinery and Intelligence", *Mind*, No. 49, pp. 433–460, 1950, available at: https://www.csee.umbc.edu/courses/471/papers/turing.pdf.

3 Shannon C., "Programming a Computer for Playing Chess", 1950, available at: https://vision.unipv.it/IA1/aa2009-2010/ProgrammingaComputerforPlayingChess.pdf.

Milestones in AI history	Year	Country
The Perceptron, designed by Frank Rosenblatt (computer model of perception by the brain)	1957	United States
Andrei Nikolaevich Kolmogorov and Vladimir Igorevich Arnold demonstrated the mathematical feasibility of neural networks	1957	USSR
Creation of the MIT Artificial Intelligence Laboratory (by John McCarthy and Marvin Minsky)	1958	United States
Development of the programming language LISP (LISt Processing)	1958	United States
End of the Cuban Revolution	1959	Cuba
Creation of the MIT Artificial Intelligence Laboratory by Marvin Minsky	1959	United States
Introduction of the term "machine learning"	1959	United States
Article by Marvin Minky, "Steps toward artificial intelligence"[4]	1961	United States
First industrial robot, UNIMATE (used at General Motors on its assembly lines)	1961	United States
Cuban Missile Crisis (October)	1962	Cuba
Creation of SAIL (Stanford Artificial Intelligence Laboratory) at the initiative of John McCarthy	1963	United States
E.A. Feigenbaum, J. Feldman, *Computers and Thought* [FEI 63]	1963	United States
Development of the STUDENT application (solves algebraic problems)	1964	United States
ELIZA, first chatbot made at MIT by Joseph Weizenbaum. Supposed to be an interactive program that allows you to converse in English	1964	United States
Lofti Zadeh[5] formalized fuzzy logic[6]	1965	United States
DENDRAL, the first expert system for chemical analysis, developed by Edward Feigenbaum and geneticist Joshua Lederberg (Stanford University)	1965	United States

4 Minsky M., "Steps toward artificial intelligence", *Proceedings of the IRE*, January 1961, pp. 8–30, available at: https://courses.csail.mit.edu/6.803/pdf/steps.pdf.

5 Lofti Zadeh is a mathematician who was born in Azerbaijan, graduated from the University of Tehran and continued his studies at MIT.

6 Fuzzy logic is a tool of AI.

Milestones in AI history	Year	Country
SHAKEY robot (Stanford) developed by a team formed around Charles Rosen	1966	United States
Terry Winograd developed the natural language application SHRDLU	1968	United States
2001: A Space Odyssey, a film in which humans are confronted with an AI, HAL 9,000	1968	United States
Mansfield amendment. DARPA will henceforth only fund projects directly related to military applications[7]	1970	United States
Presentation of the first facial recognition system, by researchers from Kyoto University (on the occasion of Osaka Expo)	1970	Japan
Masahiro Mori publishes "The Uncanny Valley". This article argues that there is an area of discomfort for humans when faced with robots	1970	Japan
Program language PROLOG	1972	United States
WABOT-1, the first "intelligent" humanoid robot	1972	Japan
James Lighthill (mathematician) told the British Science Council that AI research had so far failed to produce any of the promised impacts	1973	UK
End of the Vietnam War	1975	Vietnam
Star Wars, film by George Lucas: humanoid robot C-3PO and droid R2-D2	1977	United States
First international conference on the applications of artificial intelligence to law (Swansea, Wales), September 17–27, 1979	1979	Wales
The Stanford Cart mobile robot developed in 1961 was improved, equipped with cameras, and became one of the first examples of an autonomous vehicle	1979	United States
Gammanoid robot beat world backgammon champion (Carnegie-Mellon University)	1979	United States/ Monte Carlo

7 Communities of researchers that had been able to form thanks to this funding then broke up and had to find new spaces to federate. Such was, for example, the role of the Palo Alto Research Center (PARC) and the Xerox company. See https://monoskop.org/images/c/c0/DeLanda_Manual_War_in_the_Age_of_Intelligent_Machines.pdf.

Milestones in AI history	Year	Country
Soviet troops invade Afghanistan (December 27, 1979)	1979	Afghanistan
WABOT-2 (Waseda University), improved version of the 1972 humanoid robot	1980	Japan
Stanford's AI Laboratory, SAIL, was merged with the Computer Science Department	1980	United States
First issue of *AI Magazine* (published by the American Association for Artificial Intelligence)[8]	1980	United States
Fifth Generation Computer Project. The government invested heavily, through its Ministry of Foreign Trade, in computer R&D, with the aim of creating computers capable of conversing, translating, interpreting images, reasoning like a human being (October 1981). The project was launched in 1982	1981	Japan
CIA Strategic Plan 1983–1993	1983	United States
Publication of the "Strategic Computing Initiative" (DARPA). The Pentagon said it wanted to develop completely autonomous weapons. DARPA planned to invest 1 billion dollars from 1983 to 1993 for the development of military applications of AI. The poor results of the first years led to the the program being abandoned fairly quickly	1983	United States
Roger Schank and Marvin Minsky predicted an AI winter would occur soon	1984	United States
Creation of the Institute for New Generation Computer Technology (ICOT)	1985	Japan
Foundation of the Japanese Society for Artificial Intelligence (JSAI)	1986	Japan
Mercedes-Benz designed a driverless vehicle	1986	Germany
Rollo Carpenter developed the chatbot Jabberwacky	1988	UK
Withdrawal of Soviet forces from Afghanistan (February 15, 1989)	1989	Afghanistan

8 All issues since 1980 are available on the magazine's website: https://www.aaai.org/ojs/index.php/aimagazine/issue/archive. This resource was used in our work, which was useful not only for reading about the development of AI over a long period of time, but also for more specific research on the discourses of the AI research community on defense applications.

Milestones in AI history	Year	Country
Beginning of the Gulf War	1990	Iraq
Creation of the Pacific Rim International Conference on Artificial Intelligence (PRICAI)	1990	Japan
Creation of the Loebner[9] Prize, an annual competition for the programs that best pass the Turing test	1990	United States
MIT robotics experts found the iRobot Corporation[10]	1990	United States
End of the Gulf War	1991	Iraq
Dissolution of the USSR (December 26, 1991)	1991	USSR
Start of the Chechen War	1993	Chechnya
Vernor Vinge developed the theme of singularity in "Technological Singularity"	1993	United States
In June 1993, a group of Japanese researchers created a robotics competition, the Robot J-League. Given the success of the initiative and the very numerous international requests, the project was modified to create the RoboCup. The project was publicly announced in September 1993	1993	Japan
Start of the First Chechen War (December 11, 1994)	1994	Chechnya
Richard Wallace developed the chatbot ALICE (Artificial Linguistic Internet Computer Entity)[11]	1995	United States
End of the First Chechen War (August 31, 1996)	1996	Chechnya
Start of the Afghan Civil War	1996	Afghanistan

9 http://www.aisb.org.uk/events/loebner-prize. Participants must propose applications that meet the Turing test. The efforts made in AI since 1990 have not yet led to the validation of the test. In conclusion of the 2018 edition, the jury was quoted: "None of the chatbots competing in the finals managed to fool a judge into believing it was human. The judges ranked the chatbots according to how human-like they were."

10 https://www.irobot.fr/about-irobot/company-information/history.

11 Test the chatbot here: https://www.pandorabots.com/pandora/talk?botid=b8d616e35e36e881.

Milestones in AI history	Year	Country
First edition of the RoboCup (Robot World Cup Initiative), international robotics tournament (Nagoya)[12]. Challenge: to beat a human football team by 2050[13]	1997	Japan
The DEEP BLUE (IBM) chess computer won over Garry Kasparov (May 1997)	1997	United States
The autonomous robot Sojourner was deployed on the surface of Mars after the successful landing of the station sent by NASA (July 4, 1997)	1997	United States
Intelligent robot with emotions, KISmet, developed at MIT	1998	United States
AIBO (Artificial Intelligence roBOt), a dog-like robot. Developed by Sony. The robot was designed to learn from its interactions with the environment, humans and other AIBO	1999	Japan
Start of the Second Chechen War (August 1999)	1999	Chechnya
Kismet robot (designed by Cynthia Breazeal) recognizing and simulating facial expressions	2000	United States
Humanoid robot ASIMO (Honda company)	2000	Japan
End of the Afghan Civil War	2001	Afghanistan
Attacks on the World Trade Center (September 11, 2001)	2001	United States
A.I. Artificial Intelligence, a film in which the central character is a humanoid child robot	2001	United States
First industrial production of autonomous robotic vacuum cleaners (iRobot Corporation)	2002	United States

12 Immediate international success with 40 participating teams, and public success with 5,000 spectators.

13 However, Alan Mackworth was credited with the idea of robot football players: “A long term goal is to have teams of robots engaged in cooperative and competitive behaviour. In particular, we have chosen soccer playing as one of the tasks.” Mackworth A., *On Seeing Robots: Computer Vision: System, Theory*, and https://pdfs.semanticscholar, pp. 1–13, World Scientific Press, Singapore, 1993, available at: Applications.org/d69a/25e87e905f55e8669bf67079faac8b8510ef.pdf. Gaming robots were developed by Alan Mackworth and Dinesh Pai (University of British Columbia) as part of the Dynamo (Dynamics and Mobile Robots) project. The idea of robotic football players was also reportedly discussed at the *Workshop on Grand Challenges in Artificial Intelligence* in October 1992 in Tokyo.

Milestones in AI history	Year	Country
Start of the Iraq War	2003	Iraq
End of the Chechen War	2004	Chechnya
Reopening of the famous Stanford AI Laboratory, SAIL (created in 1963 and then merged with the Computer Science Department in 1980)	2004	United States
NASA robots (Spirit and Opportunity) explored the soil of Mars	2004	United States
Introduction of the term "machine reading" (by three researchers, Oren Etzioni, Michele Banko and Michael Cafarella) (refers to unsupervised independent text comprehension)	2006	United States
Start of Operation Enduring Freedom	2007	Sahara
Russian–Georgian conflict	2008	Georgia
Google was working on the development of an autonomous car	2009	United States
End of the Second Chechen War (April 2009)	2009	Chechnya
ImageNet created the ImageNet Large Scale Visual Recognition Challenge (ILSVRC), an object recognition competition	2010	United States
Beginning of the Arab Spring	2010	Arab States
End of the Iraq War	2011	Iraq
Libyan Civil War	2011	Libya
Start of the Syrian Civil War	2011	Syria
Start of the Google Brain Project	2011	United States
SIRI, intelligent virtual assistant, is integrated into the iPhone 4S mobile phone	2011	United States
IBM's WATSON application won a *Jeopardy* question/answer game	2011	United States
NASA sent a humanoid robot, Robonaut, developed in collaboration with General Motors, into space	2011	United States
End of the Arab Spring	2012	Arab States
Two Google researchers trained a network of 16,000 processors to recognize images of cats, showing them 10 million images of YouTube videos	2012	United States
A neural network won the ImageNet challenge	2012	Canada

Milestones in AI history	Year	Country
Never Ending Image Learner (NEIL), a machine learning system that compares and analyzes relationships between images (Carnegie Mellon University)	2013	United States
IBM's Watson application goes to market	2013	United States
Launch of the *Human Brain Project* (HBP) by the European Commission	2013	European Union
Japan sends a robot called Kirobo (developed by Toyota) to the ISS (International Space Station)	2013	Japan
Cortana, Microsoft's virtual assistant	2014	United States
Eugene Goostman chatbot passed the Turing test	2014	Russia
ALEXA, Amazon's intelligent virtual assistant (for intelligent speakers)	2014	United States
Petition signed by 3,000 people around the world, calling for a ban on automatic weapons	2015	International
Google DeepMind's AlphaGo beat the European Gogame champion (October 2015)	2015	United States
The robot Claudico played poker against four of the best players in the world and lost the games. His successor, Libratus, won the games in 2017	2015	United States
Google released its Deep Learning software, TensorFlow	2015	United States
ALphaGo (Google DeepMind) beat South Korean Go player Lee Sedol (by four wins for the machine against one for the man)[14] (March 2016)	2016	United States
Microsoft's TAY chatbot spread hate on social networks	2016	United States
Google Home, speakerphone with built-in AI to act as a personal assistant	2016	United States
Hanson Robotics created the Sophia robot	2016	United States
An improved version of AlphaGo (Google DeepMind) won 60 victories (out of 60 games) against the world's best Go players (January 4, 2017)	2017	United States

14 This victory of the machine over human is called the "Sputnik moment", to designate a high point, a historical moment, as well as one with strategic implications. https://www.dw.com/en/whos-afraid-of-artificial-intelligence-in-china/a-45546972.

Milestones in AI history	Year	Country
Saudi Arabia was the first country in the world to grant citizenship to a humanoid robot named Sophia (developed by Hanson Robotics) on October 26, 2017	2017	Saudi Arabia
The Facebook Artificial Intelligence Research lab trained two chatbots to chat. They abandoned the English language and "created" a language of their own	2017	United States
The AI Libratus program played poker against four of the world's best players for 20 days and won (Carnegie Mellon University)	2017	United States
Alibaba's language processing AI outperformed humans in a Stanford test on reading and comprehension of text	2018	China
Bixby, Samsung's virtual assistant	2018	South Korea
A set of five neural networks beat professional gamers in *Dota 2* video game	2018	United States
Russia sent a humanoid robot[15] – Skybot F850 – to the ISS[16] (August 22, 2019)	2019	Russia
Publication of the summary of the U.S. DoD AI strategy, "2018 Department of Defense Artificial Intelligence Strategy. Harnessing AI to Advance Our Security and Prosperity"[17] (February 12, 2019)	2019	United States

Table A1.1. *A chronology of AI dates and milestones*

15 Height: 6 feet. Weight: 160 kg.

16 The objective of sending robots into space is to make them carry out missions that are dangerous for humans and to make them assistants or actors in the conquest of space, in a distant space. See similar initiatives by NASA in 2011 and Japan in 2013.

17 https://media.defense.gov/2019/Feb/12/2002088963/-1/-1/1/SUMMARY-OF-DOD-AI-STRATEGY.PDF.

Appendix 2

AI in *Joint Publications* (Department of Defense, United States)

Document	Year	Artificial intelligence	Machine learning	Deep learning	Autonomous	Expert systems
"Reconnaissance, Surveillance, and Target Acq Sppt for Joint Op. JP 3-55"	1993	0	0	0	1[1]	0
"JTTP for Unmanned Aerial Vehicles. JP 3-55.1"	1993	0	0	0	4	0
"Close Air Support. JP 3-09.3"	2014	0	0	0	0[2]	0
"Counterterrorism. JP 3-26"	2014	0	0	0	0[3]	0
"Joint Airspace Control. JP 3-52"	2014	0	0	0	2	0
"Barriers, Obstacles, and Mine Warfare for Joint Operations. JP 3-15"	2016	0	0	0	2	0

1 An unmanned aerial vehicle is a powered, aerial vehicle that does not carry a human operator, uses aerodynamic forces to provide vehicle lift, can fly autonomously or be piloted remotely.

2 The notion appears 14 times in the document, but does not refer to autonomous systems in the sense of intelligent machines or autonomous robots.

3 Here, autonomy refers to that of terrorist action.

Document	Year	Artificial intelligence	Machine Learning	Deep learning	Autonomous	Expert systems
"Joint and National Intelligence Support to Military Operations. JP 2-01"	2017	1	0	0	0	0
"Countering Air and Missile Threats. JP 3-01"	2017	0	0	0	0[4]	0
"Counterinsurgency. JP 3-24"	2018	0	0	0	0[5]	0
"Meteorological and Oceanographic Operations. JP 3-59"	2018	0	0	0	1[6]	0
"Joint Logistics. JP 4-0"	2019	1	1	0	0	0

Table A2.1. *The place of AI in the U.S. military doctrinal corpus (Joint Publications)*

4 In this document, the term "autonomous" appears 10 times, but does not refer to autonomous machines or systems. It refers to the autonomous decisions of individuals.

5 As in other documents in this list, the term "autonomous" that appears refers to the actions of groups and individuals, not to that of intelligent machines.

6 "The Navy acquires […] remotely sensed data from airborne and satellite sensors, buoys, and autonomous underwater vehicles and gliders." https://fas.org/irp/doddir/dod/jp3_59.pdf.

Appendix 3

AI in the Guidelines and Instructions of the Department of Defense (United States)

Title of the document	Year	Artificial intelligence	Machine learning	Deep learning	Autonomous	Expert system
"DoDD 3000.09 Autonomy in Weapon Systems, November 21, 2012, Incorporating Change 1, May 8, 2017"	2017	0	0	0	108	0

Table A3.1. *AI and its associated concepts, search in DoD documents dedicated to "cyber" or "operations information" issues (directives, instructions and the like)*[1]

1 A long list of DoD guidelines and instructions can be found at: https://fas.org/irp/doddir/dod/index.html. From this large body of work, we have retained only those texts whose titles indicate that they deal primarily or solely with cyberspace, cyber defense, and information and communication technology issues, under the headings of "information operations" and "information warfare". The objective is to observe the presence – or not – of "artificial intelligence" in these cyber-centered discourses.

Appendix 4

AI in U.S. Navy Instructions

All the documents listed in Table A4.1 were gathered online[1].

Document	Year	Artificial intelligence	Machine learning	Deep learning	Autonomous	Expert systems
"OPNAV INSTRUCTION 5513.1F. Department of the Navy Security Classification Guides"	2005	3	0	0	0	0
"OPNAV INSTRUCTION 5513.16B"	2006	0	0	0	2[2]	0
"OPNAV INSTRUCTION 5513.7D"	2008	0	0	0	1[3]	0
"OPNAV INSTRUCTION 5513.15E"	2008	0	0	0	1[4]	0

Table A4.1. *AI in U.S. Navy documents*

1 https://fas.org/irp/doddir/navy/index.html.

2 Battlespace Preparation Autonomous Undersea Vehicle (BPAUV); Semi-Autonomous Hydrographic Reconnaissance Vehicle (SAHRV).

3 Reference to autonomous vehicles: "Battlespace Preparation Autonomous Undersea Vehicle".

4 Semi-Autonomous Hydrographic Reconnaissance Vehicle (SAHRV).

The SecNavInst instruction set is available online[5].

All OPN Instructions (Chief of Naval Operations Instructions Navy Intelligence and Security Doctrine) are available online[6].

5 https://fas.org/irp/doddir/navy/secnavinst/index.html.

6 https://fas.org/irp/doddir/navy/opnavinst/index.html.

Appendix 5

AI in U.S. Marine Corps Documents

All of the documents listed in Table A5.1 were gathered online[1].

Document	Year	Artificial intelligence	Machine learning	Deep learning	Autonomous	Expert systems
"MCDP 6 Command and Control"	1996	2	0	0	0	2
"MCWP 3-42.1. Unmanned Aerial Vehicle Operations, 14 August 2003"	2003	0	0	0	2	0
"Marine Corps Midrange Threat Estimate: 2005–2015, Marine Corps Intelligence Activity, July 1, 2005"	2005	0	0	0	0[2]	0

Table A5.1. *AI in U.S. Marine Corps documents*

1 https://fas.org/irp/doddir/usmc/index.html.

2 The term appears three times, but in reference to "autonomous regions".

Appendix 6

AI in U.S. Air Force Documents

All of the documents listed in Table A6.1 were gathered online[1].

Document	Year	Artificial intelligence	Machine learning	Deep learning	Autonomous	Expert systems
"Air Force Handbook Vol. 5 36-2235"	2002	0	0	0	0	5
"AFDD 3-13, Information Operations, January 11, 2005, incorporating change 1, July 28, 2011"	2005/2011	1	0	0	0	0
"Air Force Material Command Instruction 23-112"	2006	1	0	0	0	0
"AFDD 3-40, Counter-Chemical, Biological, Radiological and Nuclear Operations, 26 January 2007, interim change 2, 1 November 2011"	2007/2011	0	0	0	1	0
"AFDD 6-0, Command and Control, 1 June 2007, incorporating change 1, 28 July 2011"	2007/2011	0	0	0	0	1

1 https://fas.org/irp/doddir/usaf/index.html.

Document	Year	Artificial intelligence	Machine learning	Deep learning	Autonomous	Expert systems
"AFDD 3-01, Counterair Operations, 1 October 2008, interim change 2, 1 November 2011"	2008/2011	0	0	0	1[2]	0
"Air Force Manual 15-129, Vol. 2"	2011/2017	1	0	0	0	0
"Energy Horizons: United States Air Force Energy S&T Vision 2011–2026, AF/ST TR 11-01, 31 January 2012"	2012	1	0	0	16	0
"Air Force Instruction 63-101/20-101, May 9, 2017"	2017	0	0	0	1	0
"Air Force Manual 15-129, Vol. 2"	2019	1[3]	0	0	0	0

Table A6.1. *AI in U.S. Air Force documents*

2 Only one occurrence strictly concerns autonomous systems. Five others relate to autonomous operations.

3 The same reference as in document version 2017.

References

[ABU 19] ABU SALEM F.K., AL FEEL R., ELBASSUONI S. *et al.*, "FA-KES: A fake news dataset around the Syrian War", *Proceedings of the Thirteenth International AAAI Conference on Web and Social Media (ICWSM 2019)*, Munich, Germany, available at: https://www.aaai.org/ojs/index.php/ICWSM/article/download/3254/3122/, pp. 575–582, 2019.

[AHA 17] AHA D.W., COMAN A., "The AI rebellion: Changing the narrative", *AAAI Conference on Artificial Intelligence*, San Francisco, United States, available at: https://www.nrl.navy.mil/itd/aic/content/ai-rebellion-changing-narrative, 2017.

[AIF 19] AI FINLAND, Artificial intelligence is advancing at full speed, Report, available at: https://www.tekoalyaika.fi/en/reports/finland-leading-the-way-into-the-age-of-artificial-intelligence/1-artificial-intelligence-is-advancing-at-full-speed/#:~:text=Artificial%20intelligence%20means%20devices%2C%20software,task%20and%20situation%20at%20hand, June 2019.

[ALB 88] ALBANO M.C., An initial study examining the feasibility of expert system technology for command and control of supporting arms in the United States Marine Corps, Master's thesis, Naval Postgraduate School, available at: https://core.ac.uk/download/pdf/36716505.pdf, 1988.

[ALL 17] ALLEN G.C., "Project Maven brings AI to the fight against ISIS", *Bulletin of the Atomic Scientists*, available at: https://thebulletin.org/2017/12/project-maven-brings-ai-to-the-fight-against-isis/, 2017.

[AND 84] ANDRIOLE S.J., HOPPLE G.W. (eds), *Defense Applications of Artificial Intelligence*, Lexington, Toronto, 1984.

[ASH 51] ASHBY W.-R., *Les mécanismes cérébraux de l'activité intelligente*, PUF, Paris, 1951.

[ASI 41] ASIMOV I., "Liar!", *Astounding*, April–May 1941.

[BAC 87] BACK Jr. J.F., BARBONE Jr. A.F., CROCKER G.K. *et al.*, Artificial intelligence: Expert systems for corps tactical planning and other applications, Study project DTIC ADA183283, US Army War College, available at: https://archive.org/download/DTIC_ADA183283/DTIC_ADA183283.pdf, 1987.

[BAK 95] BAKER N., COSGROVE P., COULTHARD-CLARK C. *et al.*, From past to future: The Australian experience of land/air operations, Report, The Australian Defence Force, available at: https://www.army.gov.au/sites/g/files/net1846/f/1995_from_past_to_future_aust_experience_of_land_air_operations_0.pdf, 1995.

[BAU 18] BAUM S.D., "Superintelligence skepticism as a political tool", *Information*, vol. 9, no. 9, available at: https://doi.org/10.3390/info9090209; https://www.mdpi.com/2078-2489/9/9/209, p. 209, 2018.

[BÉG 50] BÉGUIN A., "L'âge des robots", *Revue Esprit*, special edition, September 1950.

[BÉN 71] BÉNARD H., BARIÉTY M., *Bulletin de l'Académie Nationale de Médecine*, Masson, Paris, available at: https://gallica.bnf.fr/ark:/12148/bpt6k63151374, 5 January 1971.

[BLO 98] BLOCH J., *La guerre : traduction de l'ouvrage russe "La guerre future aux points de vue technique, économique et politique"*, vol. 1, Guillaumin, Paris, available at: https://gallica.bnf.fr/ark:/12148/bpt6k96040616, 1898.

[BLU 22] BLUMENFELD J., "Les idées à travers le monde – R.U.R.", *Floréal : l'hebdomadaire illustré du monde du travail*, available at: https://gallica.bnf.fr/ark:/12148/bpt6k63083399/f11.image.r=%22un%20robot%22?rk=21459;2, 9 December 1922.

[BOD 77] BODEN M., *Artificial Intelligence and Natural Man*, Basic Books, Inc., New York, available at: https://profiles.nlm.nih.gov/ps/access/BBBBLN.pdf, 1977.

[BUC 76] BUCHANAN B.G., LEDERBERG J., MCCARTHY J., Three reviews of J. Weizenbaum's computer power and human reason, Research project, Stanford Artificial Intelligence Laboratory, available at: https://apps.dtic.mil/dtic/tr/fulltext/u2/a044713.pdf, 1976.

[BUC 85] BUCHANAN B.G., Expert systems: Working systems and the research literature, Report, Stamford University, available at: http://infolab.stanford.edu/pub/cstr/reports/cs/tr/85/1075/CS-TR-85-1075.pdf, October 1985.

[BUR 12] BURMAN J.T., "The misunderstanding of memes: Biography of an unscientific object, 1976–1999", *Perspectives on Science*, vol. 20, no. 1, available at: https://www.mitpressjournals.org/doi/pdf/10.1162/POSC_a_00057, pp. 75–104, 2012.

[CDI 19] CENTER FOR DATA INNOVATION, RFI: Developing a federal AI standards engagement plan, Letter, available at: http://www2.datainnovation.org/2019-nist-ai-standards.pdf, 10 May 2019.

[CHA 50] CHAUCHARD P., "Les cerveaux artificiels", *Revue Esprit*, September 1950.

[CHI 19] CHIN L., Maintaining American values with AI ethics standards for data collection and algorithmic processing, Report, available at: https://www.nist.gov/system/files/documents/2019/06/03/nist-ai-rfi-lchin_001.pdf, 31 May 2019.

[CHR 01] CHRISTALLER T., DECKER M., GILSBACH J.-M. *et al.*, *Robotik – Perspektiven für menschliches Handeln in der zukünftigen Gesellschaft, Wissenschaftsethik und Technikfolgenbeurteilung Band 14*, Springer, Berlin, 2001.

[CLA 17] CLAIR S., "Akademgorodok, la cité où se forment les soldats de la cyberguerre de Poutine", *Slate*, available at: http://www.slate.fr/story/140687/bienvenue-akademgorodok-capitale-russe-cyberguerre, 15 March 2017.

[CLE 71] CLEMA J., KIRKHAM J., "CONSIM (Conflict Simulator): Risk, cost and benefit in political simulation", *Proceeding ACM'71. Proceedings of the 1971 26th Annual Conference*, New York, United States, available at: https://dl.acm.org/citation.cfm?id=810488, pp. 226–235, 1971.

[COM 17] COMAN A., JOHNSON B., BRIGGS G. *et al.*, "Social attitudes of AI rebellion: A framework", *Workshop on AI, Ethics, & Society of the 2017 AAAI Conference on Artificial Intelligence*, San Francisco, USA, available at: https://www.nrl.navy.mil/itd/aic/content/social-attitudes-ai-rebellion-framework, 2017.

[CON 18] CONGER K., CAMERON D., "Google is helping the Pentagon build AI for drones", *Gizmodo*, available at: https://gizmodo.com/google-is-helping-the-pentagon-build-ai-for-drones-1823464533, 3 June 2018.

[COS 54] COSSA P., *Du cerveau humain aux cerveaux artificiels*, Masson et Cie, Paris, 1954.

[COU 52] COUFFIGNAL L., *Les machines à penser*, Les Éditions de Minuit, Paris, 1952.

[CSL 18] CENTER FOR STRATEGIC LEADERSHIP, Strategic cyberspace operations guide, Report, United States Army War College, available at: https://csl.armywarcollege.edu/USACSL/Publications/Strategic_Cyberspace_Operations_Guide.pdf, 30 November 2018.

[CUT 95] CUTMORE T.R.H., GAMBLE H.D., An expert system opponent for wargaming, Report, Defense and Civil Institute of Environmental Medicine, available at: https://ia800105.us.archive.org/24/items/DTIC_ADA301667/DTIC_ADA301667.pdf, April 1995.

[DAN 86] DANIELS J.D., "Artificial intelligence: A brief tutorial", *Signal*, June 1986.

[DAR 83] DARPA, Strategic computing, new-generation computing technology. A strategic plan for its development and application to critical problems in defense, Report, Defense Advanced Research Projects Agency, available at: https://apps.dtic.mil/dtic/tr/fulltext/u2/a141982.pdf, 28 October 1983.

[DAR 11] DARPA, Fiscal year (FY) 2012 budget estimates, Justification dossier, Department of Defense, United States, available at: https://www.darpa.mil/attachments/(2G5)%20Global%20Nav%20-%20About%20Us%20-%20Budget%20-%20Budget%20Entries%20-%20FY2012%20(Approved).pdf, February 2011.

[DAR 18] DARKTRACE, The next paradigm shift – AI-driven cyber-attacks, White Paper, BrightTALK, available at: https://www.brighttalk.com/webcast/13279/356575/the-next-paradigm-shift-ai-driven-cyber-attacks, 2018.

[DAV 84] DAVIS P.K., Rand's experience in applying artificial intelligence techniques to strategic-level military-political war gaming, Document, Rand Corporation, available at: https://ia800101.us.archive.org/13/items/DTIC_ADA147272/DTIC_ADA147272.pdf, 1984.

[DAW 76] DAWKINS R., *The Selfish Gene*, Oxford University Press, Oxford, 1976.

[DDI 84] DDI, Transcript of DDI Remarks to the security affairs support association, Transcription, CIA, available at: https://www.cia.gov/library/readingroom/docs/CIA-RDP95M00249R000801140011-6.pdf, 13 November 1984.

[DEL 53] DE LATIL P., *Introduction à la cybernétique. La pensée artificielle*, Gallimard, Paris, available at: https://gallica.bnf.fr/ark:/12148/bpt6k3403312q, pp. 206–227, 1953.

[DEL 91] DE LANDA M., *War in the Age of Intelligent Machines*, MIT Press, New York, available at: https://monoskop.org/images/c/c0/DeLanda_Manuel_War_in_the_Age_of_Intelligent_Machines.pdf, 1991.

[DES 85] DESBOIS D., "Comment l'intelligence artificielle conduirait la guerre", *Le Monde Diplomatique*, available at: https://www.monde-diplomatique.fr/1985/09/DESBOIS/38774, pp. 19–21, September 1985.

[DNI 19] DIRECTOR OF NATIONAL INTELLIGENCE, The AIM initiative, a strategy for augmenting intelligence using machines, Document, available at: https://www.dni.gov/files/ODNI/documents/AIM-Strategy.pdf, 2019.

[DOD 90] DEPARTMENT OF DEFENSE, Critical technologies plan, Report, available at: https://ia800105.us.archive.org/25/items/DTIC_ADA219300/DTIC_ADA219300.pdf, 15 March 1990.

[DOD 12] DEPARTMENT OF DEFENSE, Autonomy in weapon systems, Order no. 3000.09, available at: https://fas.org/irp/doddir/dod/d3000_09.pdf, 21 November 2012.

[DOD 18a] DEPARTMENT OF DEFENSE, Summary of the 2018 National Defense Strategy of the United States of America, Document, available at: https://dod.defense.gov/Portals/1/Documents/pubs/2018-National-Defense-Strategy-Summary.pdf, 2018.

[DOD 18b] DEPARTMENT OF DEFENSE, Summary of the Cyber Strategy 2018, Document, available at: https://media.defense.gov/2018/Sep/18/2002041658/-1/-1/1/CYBER_STRATEGY_SUMMARY_FINAL.PDF, September 2018.

[DOD 19] DEPARTMENT OF DEFENSE, Summary of the 2018 Department of Defense Artificial Intelligence Strategy, Document, available at: https://media.defense.gov/2019/Feb/12/2002088963/-1/-1/1/SUMMARY-OF-DOD-AI-STRATEGY.PDF, February 2019.

[DON 77] DEPARTMENT OF THE NAVY/OFFICE OF NAVAL RESEARCH TOKYO, *Scientific Bulletin*, vol. 2, no. 3, available at: https://ia801908.us.archive.org/9/items/DTIC_ADA045420/DTIC_ADA045420.pdf, June 1977.

[DON 86] DONNELLY W.F., Weekly report for period ending 24 October 1986, Report, available at: https://www.cia.gov/library/readingroom/docs/CIA-RDP89-00063R000200220007-5.pdf, October 1986.

[DON 96] DEPARTMENT OF THE NAVY, Command and Control, US Marine Corps MCDP 6, Document, available at: https://fas.org/irp/doddir/usmc/mcdp6.pdf, October 1996.

[DON 05] DEPARTMENT OF THE NAVY, Department of the Navy Security Classification Guides, Guide, available at: https://fas.org/irp/doddir/navy/opnavinst/5513_1f.pdf, 7 December 2005.

[DRE 72] DREYFUS H.L., *What Computers Can't Do*, Harper and Row, New York, 1972.

[DSB 94] DEFENSE SCIENCE BOARD, Report of the Defense Science Board summer study task force on information architecture for the battlefield, Report, OATSD, October 1994.

[DSD 17] DEPUTY SECRETARY OF DEFENSE, Establishment of an algorithmic warfare cross-functional team (Project Maven), Memorandum, available at: https://www.govexec.com/media/gbc/docs/pdfs_edit/establishment_of_the_awcft_project_maven.pdf, 26 April 2017.

[DSD 18] DEPUTY SECRETARY OF DEFENSE, Establishment of the joint artificial intelligence center, Memorandum, available at: https://admin.govexec.com/media/establishment_of_the_joint_artificial_intelligence_center_osd008412-18_r....pdf, 27 June 2018.

[DTE 86] DIRECTOR OF TRAINING AND EDUCATION, Carnegie-Mellon University, Memorendum, available at: https://www.cia.gov/library/readingroom/docs/CIA-RDP88G01332R000300300014-1.pdf, 2 September 1986.

[DUB 48] DUBARLE D., "Une nouvelle science : la cybernétique – vers la machine à gouverner", *Le Monde*, 28 December 1948.

[EAR 73] EARNEST L., The first ten years of artificial intelligence research at Stanford, Report, ARPA, available at: http://i.stanford.edu/pub/cstr/reports/cs/tr/74/409/CS-TR-74-409.pdf, July 1973.

[EAT 18] EATON J., "An emerging capability: Military applications of artificial intelligence and machine learning", *Small War Journals*, available at: https://smallwarsjournal.com/jrnl/art/emerging-capability-military-applications-artificial-intelligence-and-machine-learning, February 2018.

[ECK 83] ECKMAN P.K., Appreciation for participation in AI Symposium, Memorandum, Deputy Director of Central Intelligence, available at: https://www.cia.gov/library/readingroom/docs/CIA-RDP85M00364R000500790001-3.pdf, 20 December 1983.

[ECK 84] ECKMAN P.K., 1983 AI Symposium Summary Report, Report, CIA, available at: https://www.cia.gov/library/readingroom/docs/CIA-RDP86M00886R000500040010-5.pdf, 12 July 1984.

[ELL 68] ELLIS E.S., *The Steam Man of the Prairies*, The American News Co., New York, 1868.

[ELL 12] ELLIOTT S., HALL J., SMITH M. *et al.*, Connected robots, White book, Atos, available at: https://atos.net/wp-content/uploads/2016/06/atos-white-paper-connected-robots.pdf, November 2012.

[ESP 18] ESPER M.T., Army Directive 2018-18 (Army Artificial Intelligence Task Force in Support of the Department of Defense Joint Artificial Intelligence Center), Directive, Secretary of the Army, available at: https://armypubs.army.mil/epubs/DR_pubs/DR_a/pdf/web/ARN13011_AD2018_18_Final.pdf, 2 October 2018.

[EYK 18] EYKHOLT K., EVTIMOV I., FERNANDES E. *et al.*, Robust physical-world attacks on deep learning visual classification, Report, CVPR, available at: https://arxiv.org/pdf/1707.08945.pdf, 2018.

[FEI 60] FEIGENBAUM E.A., "Soviet cybernetics and computer sciences", *Communications of the ACM*, vol. 4, no. 12, available at: https://pdfs.semanticscholar.org/7c24/c0a387c9584cfa8255b0c0a6cf7e94a84459.pdf, pp. 566–579, 1960.

[FEI 63] FEIGENBAUM E., FELDMAN J., *Computers and Thought*, McGraw-Hill, New York, 1963.

[FEI 80] FEIGENBAUM E.A., Expert systems in the 1980s, Report, Stanford University, available at: https://stacks.stanford.edu/file/druid:vf069sz9374/vf069sz9374.pdf, 1980.

[FIN 11] FINKELSTEIN R., "Tutorial: Military memetics", *Social Media for Defense Summit*, Alexandria, United States, available at: https://ia800407.us.archive.org/17/items/MilitaryMemetics/MilitaryMemetics.pdf, 24–26 October 2011.

[FIT 83] FITZWATER H.E., Development of CA Strategic Plan 1983–1993, Memorandum, available at: https://www.cia.gov/library/readingroom/docs/CIA-RDP85B01152R000901240007-7.pdf, 16 December 1983.

[FRA 86] FRANKLIN J., SHUMAKER R.P., "Artificial intelligence in military applications", *Signal*, June 1986.

[FRE 17] FREEDBERG S.J., "Artificial Intelligence will help hunt Daesh by December", *Breaking Defense*, available at: https://breakingdefense.com/2017/07/artificial-intelligence-will-help-hunt-daesh-by-december/, 13 July 2017.

[FUN 14] FUNK M., IRRGANG B. (eds), *Robotics in Germany and Japan*, Peter Lang, Francfort-sur-le-Main, available at: https://www.peterlang.com/view/9783653999648/xhtml/chapter007.xhtml, 2014.

[GAL 18] GALANINA A., "Armes intelligentes. La voie russe de l'intelligence artificielle militaire", *Izvetsia*, available at: https://iz.ru/815370/angelina-galanina-dmitrii-liudmirskii-roman-kretcul/oruzhie-razuma-rossiiskii-put-k-voennomu-iskusstvennomu-intellektu, 22 November 2018.

[GIF 84] GIFFRAIN R.J., "Qui a peur de l'intelligence artificielle ?", *Revue des Deux Mondes*, pp. 594–596, December 1984.

[GIL 85] GILHOOL G.T., A one-day "executives-only" strategy briefing, Temple University, Archives, CIA, available at: https://www.cia.gov/library/readingroom/docs/CIA-RDP87M00539R000600730002-1.pdf, 30 April 1985.

[GOF 06] GOFFI J.-Y., *Regards sur les technosciences*, Vrin, Paris, 2006.

[HAG 14] HAGEL C., Secretary of Defense Speech, Ronald Reagan Presidential Library, Simi Valley, available at: https://www.defense.gov/Newsroom/Speeches/Speech/Article/606635/, 15 November 2014.

[HAN 92] HANSON R.L., The evolution of artificial intelligence and expert computer systems in the army, Master's thesis, Faculty of the U.S. Army Command and General Staff College, Fort Leavenworth, available at: https://apps.dtic.mil/dtic/tr/fulltext/u2/a255221.pdf, 1992.

[HAN 10] HANCOCK B.J., "Memetic warfare: The future of war", *Military Intelligence Professional Bulletin*, available at: https://fas.org/irp/agency/army/mipb/2010_02.pdf, pp. 41–46, March–April 2010.

[HAO 19] HAO K., "AI is sending people to jail – and getting it wrong", *Technology Review*, available at: https://www.technologyreview.com/s/612775/algorithms-criminal-justice-ai/, 21 January 2019.

[HAR 88] HARRISON D.F., "Computers, electronic data, and the Vietnam War", *Archivaria*, no. 26, available at: https://archivaria.ca/index.php/archivaria/article/viewFile/11490/12434, pp. 18–32, 1988.

[HAR 89] HARKEN H.D., An expert system for automating nuclear strike aircraft replacement, aircraft beddown, and logistics movement for the theater warfare exercise, Master's thesis, Air Force Institute of Technology, Wright-Patterson Air Force Base, available at: https://ia802800.us.archive.org/10/items/DTIC_ADA215728/DTIC_ADA215728.pdf, December 1989.

[HOL 76] HOLDEN A.D.C., "Trends in artificial intelligence", *IEEE Transactions on Computers*, vol. C25, no. 4, available at: https://www.computer.org/csdl/trans/tc/1976/04/01674610.pdf, pp. 313–316, April 1976.

[HOR 90] HORVITZ E., "Automated reasoning for biology and medicine. Invited opening talk", *Conference on AI in Systematic Biology*, Napa Valley, United States, September 1990.

[JCS 13] JOINT CHIEFS OF STAFF, Cyberspace Operations, Joint Publication 3-12 (R), Department of Defense, Washington, DC, available at: https://fas.org/irp/doddir/dod/jp3_12r.pdf, 5 February 2013.

[JCS 14] JOINT CHIEFS OF STAFF, Joint Airspace Control, Joint Publication 3-52, Department of Defense, Washington, DC, available at: https://fas.org/irp/doddir/dod/jp3_52.pdf, 13 November 2014.

[JCS 16] JOINT CHIEFS OF STAFF, Barriers, Obstacles, and Mine Warfare for Joint Operations, Joint Publication 3-15, Department of Defense, Washington, DC, available at: https://fas.org/irp/doddir/dod/jp3_15.pdf, 6 September 2016.

[JCS 17] JOINT CHIEFS OF STAFF, Joint and National Intelligence Support to Military Operations, Joint Publication 2-01, Department of Defense, Washington, DC, available at: https://fas.org/irp/doddir/dod/jp2_01.pdf, 5 July 2017.

[JCS 18] JOINT CHIEFS OF STAFF, Cyberspace Operations, Joint Publication 3-12, Department of Defense, Washington, DC, available at: https://www.jcs.mil/Portals/36/Documents/Doctrine/pubs/jp3_12.pdf?ver=2018-06-19-092120-930, 8 June 2018.

[JCS 19] JOINT CHIEFS OF STAFF, Joint Logistics, Joint Publication 4-0, Department of Defense, Washington, DC, available at: https://fas.org/irp/doddir/dod/jp4_0.pdf, 4 February 2019.

[JIE 18] JIESHU W., "The early history of artificial intelligence in China (1950s–1980s)", *Graduate Student Workshop at 2018 Annual Meeting of the Society for the History of Technology (SHOT)*, St. Louis, United States, available at: http://wangjieshu.com/2018/10/17/history_of_ai_in_china/, 2018.

[JOR 94] JORION P., "L'intelligence artificielle au confluent des neurosciences et de l'informatique", *Lekton*, vol. 4, no. 2, available at: http://cogprints.org/491/1/LEKTON.html, pp. 85–114, 1994.

[JOS 97] JOSEPHSON P.R., "New Atlantis revisited: Akademgorodok, the Siberian city of science", *Faculty Books*, vol. 1, available at: https://digitalcommons.colby.edu/cgi/viewcontent.cgi?article=1000&context=facultybooks, 1997.

[KAS 17] KASHIN V., RASKA M., Countering the U.S. third offset strategy: Russian perspectives, responses and challenges, Policy report, RSIS, Singapore, available at: https://www.rsis.edu.sg/wp-content/uploads/2017/01/PR170124_Countering-the-U.S.-Third-Offset-Strategy.pdf, January 2017.

[KIR 96] KIRK D.C., Artificial intelligence applications to information warfare, Document, United States Army, U.S. Army War College, Carlisle, United States, available at: https://archive.org/details/DTIC_ADA309 400, 22 March 1996.

[KOP 65] KOPROWSKI H., BRODY J.A., HADLOW W.J. *et al.*, "A new science city in Siberia", *Science 27*, vol. 149, no. 3687, available at: http://science.sciencemag.org/content/149/3687/947.extract.jpg, pp. 947–949, August 1965.

[KOZ 18] KOZYULIN V., Trois groupes de menaces constitués par les systèmes autonomes létaux, Report, available at: https://russiancouncil.ru/analytics-and-comments/analytics/tri-gruppy-ugroz-smertonosnykh-avtonomnykh-sistem/, 1 November 2018.

[KRA 88] KRAMER B.M., "Expert systems: Techniques, tools, and applications, book reviews", *AI Magazine*, vol. 9, no. 1, available at: https://www.aaai.org/ojs/index.php/aimagazine/article/view/667/585, 1988.

[LAN 69] LANGEVIN L., "Les machines à penser", *La Pensée : revue du rationalisme moderne*, no. 147, pp. 61–89, October 1969.

[LAU 85] LAURENT J.-P., "AI research in France", *AI Magazine*, vol. 6, no. 1, available at: https://doi.org/10.1609/aimag.v6i1.465, pp. 22–30, 1985.

[LAW 84] LAWLOR J., Artificial intelligence and expert systems, Report, Southwest Regional Laboratory for Educational Research and Development, Los Alamitos, available at: https://files.eric.ed.gov/fulltext/ED250156.pdf, 15 January 1984.

[LAZ 88] LAZORTHES G., *Le cerveau et l'ordinateur : étude comparée des structures et des performances*, Privat, Toulouse, 1988.

[LED 76] LEDERBERG J., *Review of Joseph Weizenbaum's Computer Power and Human Reason*, Stanford University, Stanford, available at: https://profiles.nlm.nih.gov/ps/access/BBBBLN.pdf, 1976.

[LEV 64] LEVIEN R., MARON M.E., Cybernetics and its development in the Soviet Union, Memorandum, The Rand Corporation, available at: https://ia802902.us.archive.org/14/items/DTIC_AD0602705/DTIC_AD0602705.pdf, July 1964.

[LEW 91] LEWONOWSKI M.C., Information war, Research report, Air War College/Air University/Maxwell Air Force Base, available at: https://apps.dtic.mil/dtic/tr/fulltext/u2/a248134.pdf, 1991.

[LI 11] LI S., LONG F., WANG Y., "Probe into principle of expert system in psychological warfare", in HUANG D.S., GAN Y., GUPTA P. *et al.* (eds), *Advanced Intelligent Computing Theories and Applications. With Aspects of Artificial Intelligence. ICIC 2011*, Springer, Berlin, available at: https://link.springer.com/chapter/10.1007/978-3-642-25944-9_43, 2011.

[LID 96] LIDSKOG R., "In science we trust? On the relation between scientific knowledge, risk consciousness and public trust", *Acta Sociologica*, vol. 39, available at: http://citeseerx.ist.psu.edu/viewdoc/download?doi=10.1.1.925.8004&rep=rep1&type=pdf, pp. 31–56, 1996.

[LIE 73] LIEBERMAN H.R., "Soviet devising a computer net for state planning", *The New York Times*, available at: https://www.nytimes.com/1973/12/13/archives/soviet-devising-a-computer-net-for-state-planning-big-network-in-us.html, 13 December 1973.

[LIE 00] LIEBOWITZ J., ADYA M., "An analysis of using expert systems and intelligent agents for the virtual library project at the Naval Surface Warfare Center-Carderock Division", in BERTOT J.C., DIAMOND FLETCHER P. (eds), *World Libraries on the Information Superhighway: Preparing for the Challenges of the New Millennium*, IGI Global, Hershey, available at: https://epublications.marquette.edu/cgi/viewcontent.cgi?article=1082&context=mgmt_fac, pp. 169–188, 2000.

[LIG 73] LIGHTHILL D., Artificial intelligence: A general summary, artificial intelligence: A paper Symposiun, Pamphlet, Science Research Council, State House, London, April 1973.

[LIU 93] LIU Z.-Y., "Applied AI news", *AI Magazine*, vol. 14, no. 1, available at: https://aaai.org/ojs/index.php/aimagazine/article/view/1038, p. 88, 1993.

[LOG 16] LOGVINSKIY A.L., Technopark of Novosibirsk, Academgorodok, Report, available at: http://www.ccir.it/ccir/wp-content/uploads/2016/02/Technopark-of-Akademgorodok.pdf, 2016.

[LOS 17] LOSEV A., "L'intelligence artificielle militaire", *Journal Arsenal de la Patrie*, vol. 32, no. 6, available at: http://arsenal-otechestva.ru/article/990-, 2017.

[LU 94] LU H.-C., MON D.L., YANG C., "Expert systems for sea mine warfare", *Defence Science Journal*, vol. 44, no. 4, available at: https://pdfs.semanticscholar.org/42c9/e25a795a4dcaea7a1452fc8373ce97460928.pdf, pp. 305–315, October 1994.

[MAD 17] MADRIGAL A.C., "The computer that predicted the U.S. would win the Vietnam war", *The Atlantic*, available at: https://www.theatlantic.com/technology/archive/2017/10/the-computer-that-predicted-the-us-would-win-the-vietnam-war/542046/, 5 October 2017.

[MAK 19] MAKAROV V., "La Russie a créé des armes dotées d'intelligence artificielle", *Popmech*, available at: https://www.popmech.ru/weapon/news-465862-v-rossii-sozdano-oruzhie-s-iskusstvennym-intellektom/, 27 February 2019.

[MAL 87] MALONEY E.J., Request for training in artificial intelligence, Classified document, CIA, available at: https://www.cia.gov/library/readingroom/docs/CIA-RDP89-00955R000200240022-6.pdf, 23 March 1987.

[MAR 18] MARTINHO-TRUSWEL E., MILLER H., ASARE I.N. *et al.*, Towards an AI Strategy in Mexico, White book, available at: http://go.wizeline.com/rs/571-SRN-279/images/Towards-an-AI-strategy-in-Mexico.pdf, 2018.

[MAS 12] MASON M.T., "Creation myths. The beginnings of robotics research", *IEEE Robotics & Automation Magazine*, available at: https://www.ri.cmu.edu/pub_files/mason2012creation.pdf, pp. 72–77, June 2012.

[MAT 98] MATTHEWS L.J. (ed.), *Challenging the United States Symmetrically and Asymmetrically: Can America Be Defeated?*, U.S. Army War College, Carlisle Barracks, available at: https://www.hsdl.org/?view&did=2207, 1998.

[MAT 16] MATAKE K., "Shogi and artificial intelligence", *Discuss Japan. Japan Foreign Policy Forum*, available at: https://www.japanpolicyforum.jp/culture/pt20160516000523.html, May 2016.

[MCC 71] MCCARTHY J., SAMUEL A., FEIGENBAUM E. *et al.*, Project technical report, Report, Stanford University, available at: http://i.stanford.edu/pub/cstr/reports/cs/tr/71/209/CS-TR-71-209.pdf, March 1971.

[MCC 76] McCarthy J., "An unreasonable book", Book critique, Stanford University, available at: http://jmc.stanford.edu/artificial-intelligence/reviews/weizenbaum.pdf, 16 September 1976.

[MCI 83] McIngvale P.H., A preliminary investigation on the application of robotics to missile fire control, Technical report 80-4-8, US Army Missile Command, available at: https://archive.org/details/DTIC_ADA140369, November 1983.

[MEL 89] Melaragno J., Allen M.K., A plan for the application of artificial intelligence to DoD logistics, Report PL816R1, Logistics Management Institute, available at: http://www.dtic.mil/dtic/tr/fulltext/u2/a216066.pdf, 1989.

[MIA 17] Miailhe N., Hode C., Making the AI revolution work for everyone, Report, The Future Society, AI Initiative, available at: http://s3.amazonaws.com/arena-attachments/1799773/6b2d1c9536379be6c7d2ecdec46be3d2.pdf?1519334012, 3 June 2017.

[MIA 18] Miailhe N., Hodes C., "La troisième ère de l'intelligence artificielle", institut.veolia.org, available at: https://www.institut.veolia.org/sites/g/files/dvc2551/files/document/2018/03/Facts-AI-03_La_troisieme_ere_de_lintelligence_artificielle_-_Nicolas_Miailhe_Cyrus_Hodes.pdf, 2018.

[MIN 61] Minsky M., "Steps toward artificial intelligence", *Proceedings of the IRE*, available at: https://courses.csail.mit.edu/6.803/pdf/steps.pdf, pp. 8–30, January 1961.

[MIN 85] Minsky M., "Communication with alien intelligence", in Regis E. (ed.), *Extraterrestrials: Science and Alien Intelligence*, Cambridge University Press, Cambridge, 1985.

[MIZ 04] Mizoguchi R., "The JSAI and AI activity in Japan", *IEEE Intelligent Systems*, vol. 19, no. 2, available at: https://doi.org/10.1109/MIS.2004.1274913, pp. 66–67, March–April 2004.

[NAG 79] Nagao M., Matsuyama T., Mori H., "Structural analysis of complex aerial photographs", *IJCAI'79 Proceedings of the 6th International Joint Conference on Artificial Intelligence*, vol. 2, available at: https://dl.acm.org/citation.cfm?id=1623054, pp. 610–616, 1979.

[NAR 10] Narita H., Advances in vessel and aircraft technologies, Report, available at: https://ia902902.us.archive.org/17/items/DTIC_ADA520910/DTIC_ADA520910.pdf, 25 May 2010.

[NEL 78] Nelson T., "Computer lab", *Omni Magazine*, available at: https://archive.org/stream/OMNI197908/OMNI_1978_11#page/n37/mode/2up, 1978.

[NEL 84] NELSON D.E., A combat battle damage assessor expert system, Document, Air Force Wright Aeronautical Laboratories, Whright-Patterson Air Force Base, available at: https://ia800106.us.archive.org/17/items/DTIC_ADA148898/DTIC_ADA148898.pdf, December 1984.

[NET 16] NETWORKING AND INFORMATION TECHNOLOGY RESEARCH AND DEVELOPMENT SUBCOMMITTEE, The National artificial intelligence research and development strategic plan, National Science and Technology Council, available at: https://www.nitrd.gov/PUBS/national_ai_rd_strategic_plan.pdf, October 2016.

[NEW 81] NEWELL A., "The Scientific relevance of robotics", *AI Magazine*, vol. 2, no. 1, available at: https://doi.org/10.1609/aimag.v2i1.95, pp. 24–34, 1981.

[NG 16] NG L.S.A., "Why AI is the new electricity", *Nikkei Asian Review Online*, 27 October 2016.

[NIK 19] NIKOLAYCHUK A., L'intelligence artificielle peut-elle déclencher une guerre mondiale ?, mail.ru, available at: https://mcs.mail.ru/blog/mozhet-li-iskusstvennyj-razum-stat-prichinoj-mirovoj-vojny/, 2019.

[NIL 09] NILSSON N.J., *The Quest for Artificial Intelligence*, Cambridge University Press, Cambridge, 2009.

[NIS 12] NISHIDA T., "The best of AI in Japan – Prologue", *AI Magazine*, available at: https://pdfs.semanticscholar.org/7bb3/1d57d890f164a0d7ab9b0a1298695861f7a7.pdf, pp. 108–111, summer 2012.

[NOË 18] NOËL J.-C., "Will artificial intelligence revolutionize the art of war ?", *Politique Étrangère*, 4, winter edition, available at: https://www.cairn-int.info/abstract-E_PE_184_0159--will-artificial-intelligence.htm, pp. 159–170, 2018.

[OWE 17] OWEN-HILL A., "What's the difference between robotics and artificial intelligence?", *Robotiq*, available at: https://blog.robotiq.com/whats-the-difference-between-robotics-and-artificial-intelligence, 19 July 2017.

[PEI 18] PEIRSON V A.L., TOLUNAY E.M., Dank learning: Generating memes using deep neural networks, Stanford University, available at: https://arxiv.org/pdf/1806.04510.pdf, June 2018.

[QUA 57] QUARLES D.A., "Preparations for tracking artificial earth-satellites at the Vanguard Computing Center", *Proceeding IRE-ACM-AIEE'57 (Eastern) Papers and discussions presented at the December 9-13, 1957, eastern joint computer conference. Computers with deadlines to meet*, available at: https://dl.acm.org/citation.cfm?id=1457730, pp. 58–64, 1957.

[QUI 22] QUILLET A. (ed.), *Floréal : l'hebdomadaire illustré du monde du travail*, Paris, available at: https://gallica.bnf.fr/ark:/12148/bpt6k63083399/f11.image.r=%22un%20robot%22?rk=21459;2, 9 December 1922.

[RAP 71] RAPHAEL B., CHAITIN L.J., DUDA R.O. *et al.*, Research and applications – Artificial intelligence, Semiannual progress report, Report, Washington, DC, available at: http://citeseerx.ist.psu.edu/viewdoc/download?doi=10.1.1.462.5526&rep=rep1&type=pdf, April 1971.

[RAY 63] RAYMOND F.H., "Analogies et intelligence artificielle", *Dialectica: International Journal of Philosophy & Official Organ of the ESAP*, vol. 17, no. 2/3, pp. 203–215, September 1963.

[REP 16] REPUBLIC OF SERBIA, National Strategy for the prosecution of war crimes, Document, available at: http://www.tuzilastvorz.org.rs/upload/HomeDocument/Document__en/2016-05/p_nac_stragetija_eng.PDF, January 2016.

[RIC 01] RICHELSON J.T., Science, Technology and the CIA, A National Security Archive Electronic Briefing Book, available at: https://nsarchive2.gwu.edu//NSAEBB/NSAEBB54/index2.html, 10 September 2001.

[RID 12] RID T., MCBURNEY P., "Cyber-weapons", *The RUSI Journal*, vol. 157, no. 1, pp. 6–13, 2012.

[ROD 81] RODNEY P.G., U.S. technology transfer to the Soviet Union: A dilemma, Report, Air War College, available at: https://apps.dtic.mil/dtic/tr/fulltext/u2/a107313.pdf, 1981.

[RUD 19] RUDINSKY A., "L'intelligence artificielle dans la guerre réseau-centrique du 21ème siècle", *Oborona*, available at: http://www.oborona.ru/includes/periodics/conceptions/2011/0905/22247309/detail.shtml, 8 August 2019.

[SAF 02] SECRETARY OF THE AIR FORCE, "Information for designers of instructional systems", *Air Force Handbook*, AFH 36-2235, Washington, DC, available at: https://static.e-publishing.af.mil/production/1/af_a1/publication/afh36-2235v5/afh36-2235v5.pdf, 1 November 2002.

[SAF 12] SECRETARY OF THE AIR FORCE, Energy horizons, Document, U.S. Air Force Energy S&T Vision 2011-2026, AF/ST TR 11-01, Washington, DC, available at: https://fas.org/irp/doddir/usaf/energy.pdf, 31 January 2012.

[SAF 13] SECRETARY OF THE AIR FORCE, Air force weather operations, Manual, US Air Force Policy Directive (AFPD) 15-1, Washington, DC, available at: https://static.e-publishing.af.mil/production/1/acc/publication/afman15-129v2_accsup_i/afman15-129v2_accsup_i.pdf, 6 August 2013.

[SAF 17] SECRETARY OF THE AIR FORCE, Integrated life cycle management, Document, US Air Force Policy Directive (AFPD) 63-1/20-1, Washington, DC, available at: https:// static.e-publishing.af.mil/production/1/saf_aq/publication/afi63-101_20-101/afi63-101_2 0-101.pdf, 9 May 2017.

[SAL 76] SALEMME A.J., *Cryptolog*, vol. 3, no. 11, available at: https://www.nsa.gov/Portals/70/documents/news-features/declassified-documents/cryptologs/cryptolog_25.pdf, November 1976.

[SAV 19] SAVIN L., "АМЕРИКАНСКИЕ ФАБРИКИ ТРОЛЛЕЙ, ИСКУССТВЕННЫЙ ИНТЕЛЛЕКТ И КИБЕРВОЙНА. ЧАСТЬ 1" [Translation: American troll factories, artificial intelligence and cyberwar. Part 1.], *Geopolitica*, available at: https://www.geopolitica.ru/article/amerikanskie-fabriki-trolley-iskusstvennyy-intellekt-i-kibervoyna-chast-1, March 2019.

[SAY 19] SAYLER K.M., Artificial intelligence and national security, Report, Congressional Research Service, Washington, DC, available at: https://fas.org/sgp/crs/natsec/R45178.pdf, 30 January 2019.

[SCI 00] SCIAVICCO L., SICILIANO B., *Modelling and Control of Robot Manipulators*, 2nd ed., Springer Verlag, London, 2000.

[SHA 16] SHARIF M., "Accessorize to a crime: Real and stealthy attacks on state-of-the-art face recognition", *Proceedings CCS'16, ACM SIGSAC Conference on Computer and Communications Security*, available at: https://dl.acm.org/citation.cfm?doid=2976749.2978392, pp. 1528–1540, 2016.

[SHA 18a] SHANAHAN J., Disruption in UAS: The algorithmic warfare cross-functional team (Project MAVEN), Presentation, Department of Defense, available at: http://airpower.airforce.gov.au/APDC/media/Events-Media/RAAF%20AP%20CONF%202018/1130-1200-Shanahan-Disruption-in-UAS-The-AWCFT.pdf, 20 March 2018.

[SHA 18b] SHARMA V., "The exciting evolution of machine learning", *Vinod Sharma's Blog*, available at: https://vinodsblog.com/2018/03/11/the-exciting-evolution-of-machine-learning/?fbclid=IwAR3WUFDuTaOP78HYlFVhvAVqYe4kS2D_O37p48VIs3M11LKrqJqO8m6jm0A, 11 March 2018.

[SHN 86] SHNEIDERMAN B., PLAISANT C., *Designing the User Interface*, 2nd ed., Addison-Wesley, Boston, 1986.

[SIM 76] SIMON H.A., NEWELL A., "Computer science as empirical inquiry: Symbols and search", *Communications of the ACM*, vol. 19, no. 3, pp. 113–126, 19 March 1976.

[SIM 88] SIMMONS R.M., SWAN: An expert system with natural language interface for tactical air capability assessment, Document, Systems Research and Applications Corporation, available at: https://ia801201.us.archive.org/34/items/NASA_NTRS_Archive_19880007869/NASA_NTRS_Archive_19880007869.pdf, 1988.

[SIM 96] SIMON H.A., *The Sciences of the Artificial*, MIT Press, Cambridge, available at: https://monoskop.org/images/9/9c/Simon_Herbert_A_The_Sciences_of_the_Artificial_3rd_ed.pdf, 1996.

[SMI 14] SMITH A., ANDERSON J., AI, robotics, and the future of jobs, Report, Pew Research Center, available at: http://www.pewinternet.org/2014/08/06/future-of-jobs/, 6 August 2014.

[SOL 16] SOLON O., "Karim the AI delivers psychological support to Syrian refugees", *The Guardian*, available at: http://www2.datainnovation.org/2016-promise-of-ai.pdf, 22 March 2016.

[STE 63] STEFFERUD E., The logic theory machine: A model heuristic program, Memorandum, RAND Corporation, available at: https://history-computer.com/Library/Logic%20Theorist%20memorandum.pdf, June 1963.

[STE 86] STEVENSON C.A., Artificial intelligence and expert systems for government executives, Report, The National War College, available at: https://apps.dtic.mil/dtic/tr/fulltext/u2/a436516.pdf, February 1986.

[STE 19] STEFANOVICH D., "Intelligence artificielle et armes nucléaires", *Russian Council*, available at: https://russiancouncil.ru/analytics-and-comments/analytics/iskusstvennyy-intellekt-i-yadernoe-oruzhie/, 6 May 2019.

[STU 19] STUMBORG M., HUGHES C., RFI: Developing a federal AI standards engagement plan, Document, Center for Naval Analyses, available at: https://www.nist.gov/system/files/documents/2019/05/23/nist-ai-rfi-stumborg-hughes-ctr-for-navel-analyses-001.pdf, May 2019.

[SYC 15] SYCHEV V., "Война без участия людей Может ли новейшее вооружение убивать без команды человека?" [Translation: War without human involvement. Can the latest weapons kill without a human team?], *Meduza*, available at: https://meduza.io/feature/2015/08/01/voyna-bez-uchastiya-lyudey, 1 August 2015.

[SZE 13] SZEGEDY C., ZAREMBA W., SUTSKEVER I. *et al.*, Intriguing properties of neural networks, Document, arxiv.org, available at: https://arxiv.org/pdf/1312.6199.pdf, 2013.

[TAY 96] TAYLOR I.W., FRANK G.W., GIIRSEL M.A., An example of how expert systems could assist operations planning, Note, Department of National Defence, Canada, available at: https://ia800703.us.archive.org/32/items/DTIC_ADA640553/DTIC_ADA640553.pdf, May 1996.

[TON 66] TONGE F.M., "A view of artificial intelligence", *Proceeding ACM'66 Proceedings of the 1966 21st National Conference*, available at: https://dl.acm.org/citation.cfm?id=810717, pp. 379–382, 1966.

[TRA 17] TRADOC, The U.S. Army functional concept for Intelligence 2020–2040, Pamphlet 525-2-1, Department of the Army, available at: https://fas.org/irp/doddir/army/tp525-2-1.pdf, February 2017.

[TRA 18a] TRADOC, The U.S. Army in multidomain operations 2028, Pamphlet 525-3-1, Department of the Army, available at: https://fas.org/irp/doddir/army/tp525-3-1.pdf, 6 December 2018.

[TRA 18b] TRADOC, U.S. Army concept: Multi-domain combined arms operations at Echelons above Brigade 2025–2045, Pamphlet 525-3-8, available at: https://fas.org/irp/doddir/army/tp525-3-8.pdf, 6 December 2018.

[TSU 79] TSUGAWA S., YATABE T., HIROSE T. *et al.*, "An automobile with artificial intelligence", *Proceedings of the 6th International Joint Conference on Artificial Intelligence Volume 2*, available at: https://dl.acm.org/citation.cfm?id=1623117, pp. 893–895, 1979.

[TUR 50] TURING A.M., "Computing machinery and intelligence", *Mind*, no. 49, pp. 433–460, 1950.

[USA 10] THE UNITED STATES ARMY, Department of the Army, Cyberspace operations concept capability plan 2016–2028, TRADOC, available at: https://fas.org/irp/doddir/army/pam525-7-8.pdf, pp. 525–7–8, 22 February 2010.

[USA 18] THE UNITED STATES ARMY, Cyberspace and electronic warfare operations 2025–2040, Pamphlet, TRADOC, available at: https://fas.org/irp/doddir/army/tp525-8-6.pdf, pp. 525–8–6, January 2018.

[USM 03] U.S. MARINE CORPS, Unmanned aerial vehicle operations, Document, Department of the Navy, available at: https://fas.org/irp/doddir/usmc/mcwp3-42-1.pdf, August 2003.

[VER 19] VERNE J., *L'étonnante aventure de la mission Barsac*, Libraire Hachette, Paris, 1919.

[VIL 18] VILLANI MISSION ON ARTIFICIAL INTELLIGENCE, What is artificial intelligence ?, Document, available at: https://www.aiforhumanity.fr/pdfs/MissionVillani_WhatisAI_ENG(1)VF.pdf, March 2018.

[WAL 93] WALLACH S.P., Standard target materials for autonomous precision strike weapons, Executive Research Project S77, The Industrial College of the Armed Forces, available at: https://ia800106.us.archive.org/33/items/DTIC_ADA276575/DTIC_ADA276575.pdf, 1993.

[WAL 18] WALLACE D., "Cyber weapon reviews under international humanitarian law: A critical analysis", *Tallinn Paper*, no. 11, available at: https://ccdcoe.org/uploads/2018/10/TP-11_2018.pdf, 2018.

[WAR 80] WARNIER J.D., *L'homme face à l'intelligence artificielle*, Éditions d'Organisation, Paris, 1980.

[WAR 08] WARE W.H., RAND and the information, *RAND Corporation*, available at: https://www.rand.org/content/dam/rand/pubs/corporate_pubs/2008/RAND_CP537.pdf, 2008.

[WAR 19] WARNER B., "Tech companies are deleting evidence of war crimes", *The Atlantic*, available at: https://www.theatlantic.com/ideas/archive/2019/05/facebook-algorithms-are-making-it-harder/588931/, 8 May 2019.

[WEI 75] WEIZENBAUM J., *Computer Power and Human Reason*, W.H. Freeman Co., San Francisco, 1975.

[WES 18] WEST D.M., ALLEN J.R., How artificial intelligence is transforming the world, Report, Brookings, available at: https://www.brookings.edu/research/how-artificial-intelligence-is-transforming-the-world/, 24 April 2018.

[WHI 95] WHITE HOUSE, A National Security Strategy of Engagement and Enlargement, Document, Government Printing Office, February 1995.

[WHI 16] WHITE HOUSE, Request for Information on the Future of Artificial Intelligence, White House Office of Science and Technology Policy (OSTP), Public Responses, available at: https://obamawhitehouse.archives.gov/sites/default/files/microsites/ostp/OSTP-AI-RFI-Responses.pdf, 1 September 2016.

[WIL 83] WILLICK M.S., "Artificial intelligence: Some legal approaches and implications", *AI Magazine*, vol. 4, no. 2, available at: https://doi.org/10.1609/aimag.v4i2.392, pp. 5–16, summer 1983.

[WIM 77] WIMBLE M., "Artificial intelligence, an evolutionary idea", *Byte Magazine*, vol. 2, no. 5, available at: https://archive.org/details/byte-magazine-1977-05/page/n27, pp. 26–50, May 1977.

[XIN 83] XINSONG J., GUONING S., YU C., "AI research in the People's Republic of China: A review", *AI Magazine*, vol. 4, no. 4, available at: https://doi.org/10.1609/aimag.v4i4.417, pp. 43–48, 1983.

Index

I, K, L

M, P, R

S, T, V, W

Other titles from

iSTE

in

Computer Engineering

2020

LAFFLY Dominique
TORUS 1 – Toward an Open Resource Using Services: Cloud Computing for Environmental Data
TORUS 2 – Toward an Open Resource Using Services: Cloud Computing for Environmental Data
TORUS 3 – Toward an Open Resource Using Services: Cloud Computing for Environmental Data

LAURENT Anne, LAURENT Dominique, MADERA Cédrine
Data Lakes
(Databases and Big Data Set – Volume 2)

OULHADJ Hamouche, DAACHI Boubaker, MENASRI Riad
Metaheuristics for Robotics
(Optimization Heuristics Set – Volume 2)

SADIQUI Ali
Computer Network Security

2019

BESBES Walid, DHOUIB Diala, WASSAN Niaz, MARREKCHI Emna
Solving Transport Problems: Towards Green Logistics

CLERC Maurice
Iterative Optimizers: Difficulty Measures and Benchmarks

GHLALA Riadh
Analytic SQL in SQL Server 2014/2016

TOUNSI Wiem
Cyber-Vigilance and Digital Trust: Cyber Security in the Era of Cloud Computing and IoT

2018

ANDRO Mathieu
Digital Libraries and Crowdsourcing
(Digital Tools and Uses Set – Volume 5)

ARNALDI Bruno, GUITTON Pascal, MOREAU Guillaume
Virtual Reality and Augmented Reality: Myths and Realities

BERTHIER Thierry, TEBOUL Bruno
From Digital Traces to Algorithmic Projections

CARDON Alain
Beyond Artificial Intelligence: From Human Consciousness to Artificial Consciousness

HOMAYOUNI S. Mahdi, FONTES Dalila B.M.M.
Metaheuristics for Maritime Operations
(Optimization Heuristics Set – Volume 1)

JEANSOULIN Robert
JavaScript and Open Data

PIVERT Olivier
NoSQL Data Models: Trends and Challenges
(Databases and Big Data Set – Volume 1)

SEDKAOUI Soraya
Data Analytics and Big Data

SALEH Imad, AMMI Mehdi, SZONIECKY Samuel
Challenges of the Internet of Things: Technology, Use, Ethics
(Digital Tools and Uses Set – Volume 7)

SZONIECKY Samuel
Ecosystems Knowledge: Modeling and Analysis Method for Information and Communication
(Digital Tools and Uses Set – Volume 6)

2017

BENMAMMAR Badr
Concurrent, Real-Time and Distributed Programming in Java

HÉLIODORE Frédéric, NAKIB Amir, ISMAIL Boussaad, OUCHRAA Salma, SCHMITT Laurent
Metaheuristics for Intelligent Electrical Networks
(Metaheuristics Set – Volume 10)

MA Haiping, SIMON Dan
Evolutionary Computation with Biogeography-based Optimization
(Metaheuristics Set – Volume 8)

PÉTROWSKI Alain, BEN-HAMIDA Sana
Evolutionary Algorithms
(Metaheuristics Set – Volume 9)

PAI G A Vijayalakshmi
Metaheuristics for Portfolio Optimization
(Metaheuristics Set – Volume 11)

2016

BLUM Christian, FESTA Paola
Metaheuristics for String Problems in Bio-informatics
(Metaheuristics Set – Volume 6)

DEROUSSI Laurent
Metaheuristics for Logistics
(Metaheuristics Set – Volume 4)

DHAENENS Clarisse and JOURDAN Laetitia
Metaheuristics for Big Data
(Metaheuristics Set – Volume 5)

LABADIE Nacima, PRINS Christian, PRODHON Caroline
Metaheuristics for Vehicle Routing Problems
(Metaheuristics Set – Volume 3)

LEROY Laure
Eyestrain Reduction in Stereoscopy

LUTTON Evelyne, PERROT Nathalie, TONDA Albert
Evolutionary Algorithms for Food Science and Technology
(Metaheuristics Set – Volume 7)

MAGOULÈS Frédéric, ZHAO Hai-Xiang
Data Mining and Machine Learning in Building Energy Analysis

RIGO Michel
Advanced Graph Theory and Combinatorics

2015

BARBIER Franck, RECOUSSINE Jean-Luc
COBOL Software Modernization: From Principles to Implementation with the BLU AGE® Method

CHEN Ken
Performance Evaluation by Simulation and Analysis with Applications to Computer Networks

CLERC Maurice
Guided Randomness in Optimization
(Metaheuristics Set – Volume 1)

DURAND Nicolas, GIANAZZA David, GOTTELAND Jean-Baptiste, ALLIOT Jean-Marc
Metaheuristics for Air Traffic Management
(Metaheuristics Set – Volume 2)

MAGOULÈS Frédéric, ROUX François-Xavier, HOUZEAUX Guillaume
Parallel Scientific Computing

MUNEESAWANG Paisarn, YAMMEN Suchart
Visual Inspection Technology in the Hard Disk Drive Industry

2014

BOULANGER Jean-Louis
Formal Methods Applied to Industrial Complex Systems

BOULANGER Jean-Louis
Formal Methods Applied to Complex Systems:Implementation of the B Method

GARDI Frédéric, BENOIST Thierry, DARLAY Julien, ESTELLON Bertrand, MEGEL Romain
Mathematical Programming Solver based on Local Search

KRICHEN Saoussen, CHAOUACHI Jouhaina
Graph-related Optimization and Decision Support Systems

LARRIEU Nicolas, VARET Antoine
Rapid Prototyping of Software for Avionics Systems: Model-oriented Approaches for Complex Systems Certification

OUSSALAH Mourad Chabane
Software Architecture 1
Software Architecture 2

PASCHOS Vangelis Th
Combinatorial Optimization – 3-volume series, 2nd Edition
Concepts of Combinatorial Optimization – Volume 1, 2nd Edition
Problems and New Approaches – Volume 2, 2nd Edition
Applications of Combinatorial Optimization – Volume 3, 2nd Edition

QUESNEL Flavien
Scheduling of Large-scale Virtualized Infrastructures: Toward Cooperative Management

RIGO Michel
Formal Languages, Automata and Numeration Systems 1: Introduction to Combinatorics on Words
Formal Languages, Automata and Numeration Systems 2: Applications to Recognizability and Decidability

SAINT-DIZIER Patrick
Musical Rhetoric: Foundations and Annotation Schemes

TOUATI Sid, DE DINECHIN Benoit
Advanced Backend Optimization

2013

ANDRÉ Etienne, SOULAT Romain
The Inverse Method: Parametric Verification of Real-time Embedded Systems

BOULANGER Jean-Louis
Safety Management for Software-based Equipment

DELAHAYE Daniel, PUECHMOREL Stéphane
Modeling and Optimization of Air Traffic

FRANCOPOULO Gil
LMF — Lexical Markup Framework

GHÉDIRA Khaled
Constraint Satisfaction Problems

ROCHANGE Christine, UHRIG Sascha, SAINRAT Pascal
Time-Predictable Architectures

WAHBI Mohamed
Algorithms and Ordering Heuristics for Distributed Constraint Satisfaction Problems

ZELM Martin *et al.*
Enterprise Interoperability

2012

ARBOLEDA Hugo, ROYER Jean-Claude
Model-Driven and Software Product Line Engineering

BLANCHET Gérard, DUPOUY Bertrand
Computer Architecture

BOULANGER Jean-Louis
Industrial Use of Formal Methods: Formal Verification

BOULANGER Jean-Louis
Formal Method: Industrial Use from Model to the Code

CALVARY Gaëlle, DELOT Thierry, SÈDES Florence, TIGLI Jean-Yves
Computer Science and Ambient Intelligence

MAHOUT Vincent
Assembly Language Programming: ARM Cortex-M3 2.0: Organization, Innovation and Territory

MARLET Renaud
Program Specialization

SOTO Maria, SEVAUX Marc, ROSSI André, LAURENT Johann
Memory Allocation Problems in Embedded Systems: Optimization Methods

2011

BICHOT Charles-Edmond, SIARRY Patrick
Graph Partitioning

BOULANGER Jean-Louis
Static Analysis of Software: The Abstract Interpretation

CAFERRA Ricardo
Logic for Computer Science and Artificial Intelligence

HOMES Bernard
Fundamentals of Software Testing

KORDON Fabrice, HADDAD Serge, PAUTET Laurent, PETRUCCI Laure
Distributed Systems: Design and Algorithms

KORDON Fabrice, HADDAD Serge, PAUTET Laurent, PETRUCCI Laure
Models and Analysis in Distributed Systems

LORCA Xavier
Tree-based Graph Partitioning Constraint

TRUCHET Charlotte, ASSAYAG Gerard
Constraint Programming in Music

VICAT-BLANC PRIMET Pascale *et al.*
Computing Networks: From Cluster to Cloud Computing

2010

AUDIBERT Pierre
Mathematics for Informatics and Computer Science

BABAU Jean-Philippe *et al.*
Model Driven Engineering for Distributed Real-Time Embedded Systems

BOULANGER Jean-Louis
Safety of Computer Architectures

MONMARCHÉ Nicolas *et al.*
Artificial Ants

PANETTO Hervé, BOUDJLIDA Nacer
Interoperability for Enterprise Software and Applications 2010

SIGAUD Olivier *et al.*
Markov Decision Processes in Artificial Intelligence

SOLNON Christine
Ant Colony Optimization and Constraint Programming

AUBRUN Christophe, SIMON Daniel, SONG Ye-Qiong *et al.*
Co-design Approaches for Dependable Networked Control Systems

2009

FOURNIER Jean-Claude
Graph Theory and Applications

GUÉDON Jeanpierre
The Mojette Transform / Theory and Applications

JARD Claude, ROUX Olivier
Communicating Embedded Systems / Software and Design

LECOUTRE Christophe
Constraint Networks / Targeting Simplicity for Techniques and Algorithms

2008

BANÂTRE Michel, MARRÓN Pedro José, OLLERO Hannibal, WOLITZ Adam
Cooperating Embedded Systems and Wireless Sensor Networks

MERZ Stephan, NAVET Nicolas
Modeling and Verification of Real-time Systems

PASCHOS Vangelis Th
Combinatorial Optimization and Theoretical Computer Science: Interfaces and Perspectives

WALDNER Jean-Baptiste
Nanocomputers and Swarm Intelligence

2007

BENHAMOU Frédéric, JUSSIEN Narendra, O'SULLIVAN Barry
Trends in Constraint Programming

JUSSIEN Narendra
A TO Z OF SUDOKU

2006

BABAU Jean-Philippe *et al.*
From MDD Concepts to Experiments and Illustrations – DRES 2006

HABRIAS Henri, FRAPPIER Marc
Software Specification Methods

MURAT Cecile, PASCHOS Vangelis Th
Probabilistic Combinatorial Optimization on Graphs

PANETTO Hervé, BOUDJLIDA Nacer
Interoperability for Enterprise Software and Applications 2006 / IFAC-IFIP I-ESA'2006

2005

GÉRARD Sébastien *et al.*
Model Driven Engineering for Distributed Real Time Embedded Systems

PANETTO Hervé
Interoperability of Enterprise Software and Applications 2005